D0938787

science is beautiful

THE HUMAN BODY UNDER THE MICROSCOPE

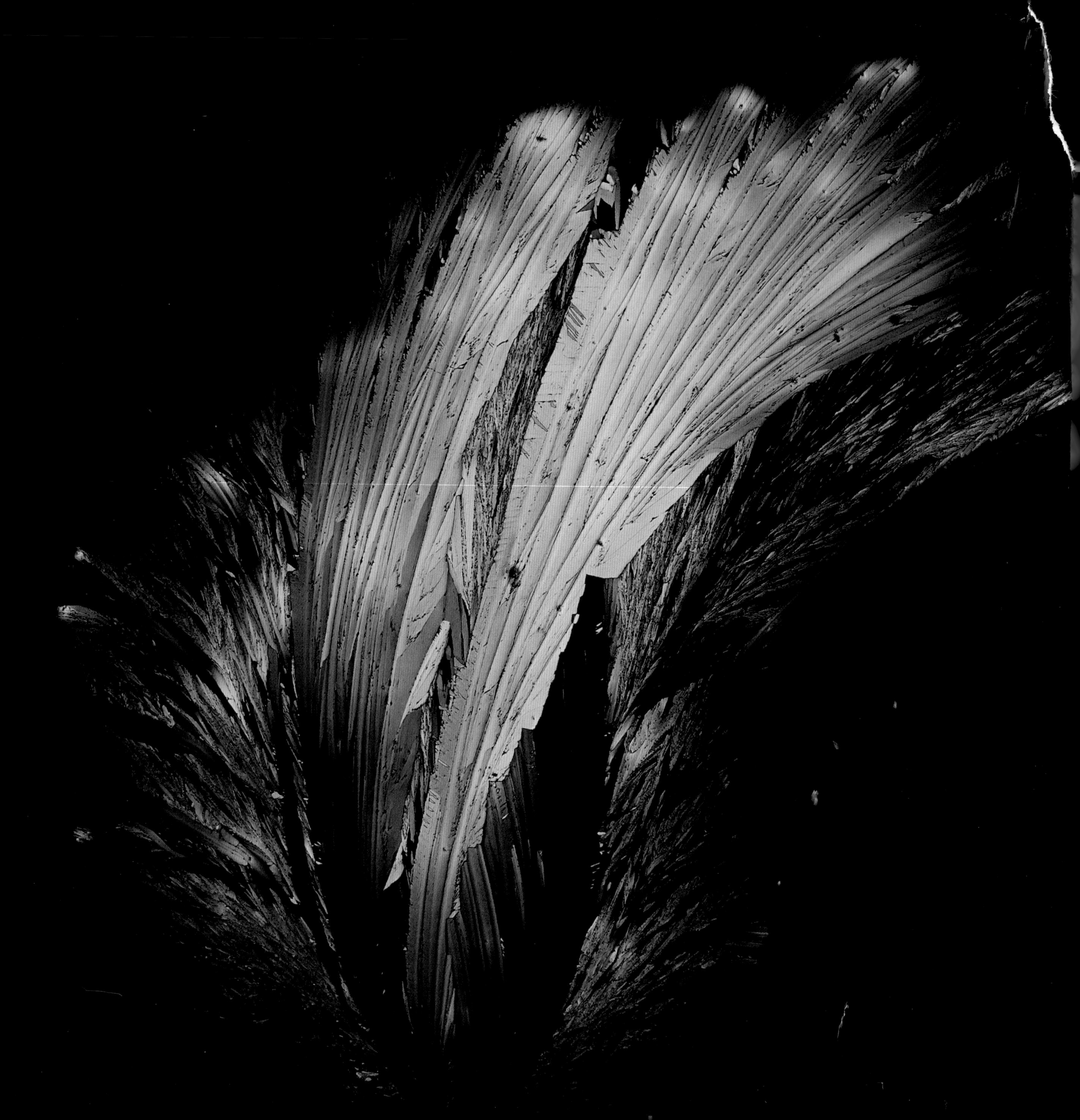

science is beautiful

THE HUMAN BODY UNDER THE MICROSCOPE

Colin Salter

BATSFORD

First published in the United Kingdom in 2014 by
Batsford, an imprint of Pavilion Books Company Limited
1 Gower Street
London WC1E 6HD
www.pavilionbooks.com

ISBN: 9781849941921

A CIP catalogue record for this book is available from the British Library.

20 19 18 17 16 15
10 9 8 7 6 5 4 3 2

Reproduction by Rival Colour Ltd, UK
Printed by 1010 Printing International Ltd, China

This book can be ordered direct from the publisher at the website:
www.pavilionbooks.com, or try your local bookshop.

Distributed in the United States and Canada by
Sterling Publishing Co., Inc. 1166 Avenue of the Americas, 17th floor
New York, NY 10036, USA

Previous page: Metformin crystals (polarized light micrograph)
These are the characteristic wing-like curves of the diabetes drug metformin: in the World Health Organisation's list of essential medicines, metformin and glyburide are the two treatments listed for oral administration for diabetes. They work in different ways: glyburide stimulates the production of insulin, which the body needs to break down blood sugars; metformin acts to reduce the production of glucose in the liver before it can enter the bloodstream.
(Magnification: x220 at 10cm/4in size)

Right: Red and white blood cells (coloured scanning electron micrograph)
Blood is the transport system for the human body. Red cells, shaped like peas squeezed between thumb and finger, contain haemoglobin which captures oxygen. The blood flows, delivering the oxygen where it is needed. The dimpled shape maximizes the oxygen-bearing area. White cells (here coloured blue) are vital to our immune system. Like brushes their tendrils trap foreign bodies and grapple with infections. Both red and white cells are generated within our bone marrow.
(Magnification: unknown)

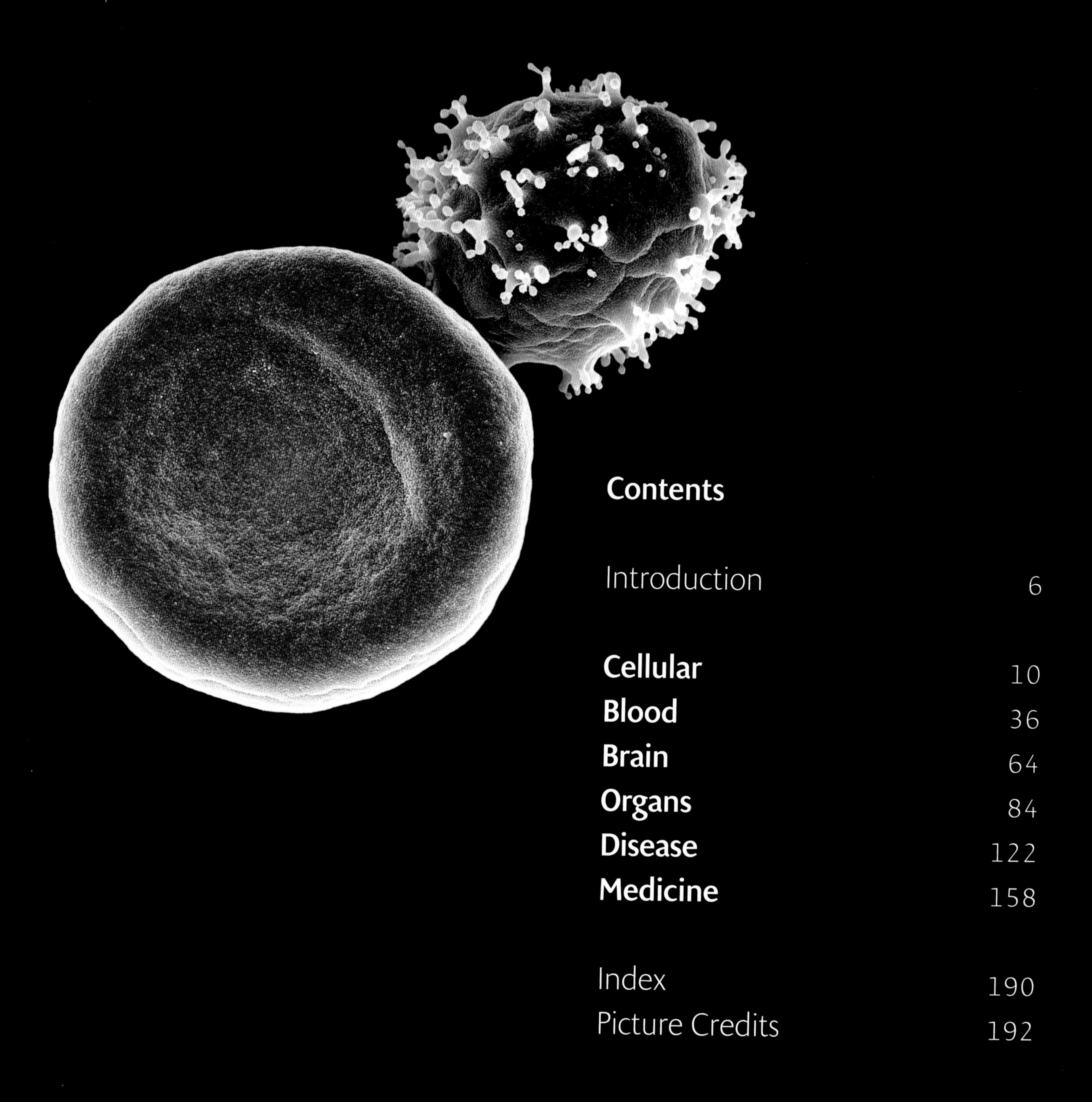

Contents

most of the time – not out of squeamishness, but because of the sheer delicate complexity of its internal systems. How, we might ask ourselves, can such an intricate and finely tuned instrument survive the knocks and scrapes of everyday life? How do such minutely engineered components cope with the wear and tear of daily use, never mind the passage of the years?

The answer is of course that they don't. They wear out, they break, they need to be repaired or replaced from time to time. Those are the jobs of our doctors and surgeons, and they are the people who need to know what we look like on the inside. But for the rest of us, understanding a little of how our bodies work is no bad thing. Often the worst thing about illness or pain, for example, is not knowing the cause of it. And often, some knowledge of the processes of our nervous, digestive and circulatory systems helps us to prevent illness and improve our health.

Medical science, like all practical sciences, is a process of observation and deduction. From the start of civilization, humans have observed which plants were medicinal for certain conditions. An Egyptian papyrus from 1600 BC about military surgery survives, proving that its unknown author had a working anatomical knowledge of the major organs of the body. Since then we have studied our internal workings in ever greater detail. The microscopic systems made visible in the photographs in this book would have been unimaginable to the ancient Egyptians, and the knowledge of ourselves which they provide is unprecedented.

So how is it possible to see inside the human body in such detail? You'll see alongside every picture in this book a note of the means by which the image was obtained. The vast majority of the pictures are micrographs. A micrograph is simply a graphic image of microscopic detail, and there are many different ways of acquiring one.

A light micrograph is produced by a light microscope. This is the traditional microscope, the one invented in the sixteenth century which uses lenses to magnify a specimen visible under natural or artificial light. When light strikes an object, it is reflected by the surfaces of that object according to the colour, texture and angle of those surfaces. That reflected light reaches the eye, either directly or (in this case) through the lenses of a light microscope. The light is gathered on light-sensitive cells inside the eye ball. The brain processes the information gathered by these cells, information about shape and size as well as colour and texture, in an activity better known to us as Sight. The light microscope sees more or less what the human eye sees, but simply magnifies it.

The microscope became a tool for scientific study in the late seventeenth century, and remains the simplest, low-tech, low-cost way to look at small things. It has changed little in essence in the four hundred years since its invention. The greatest innovations have been in the kind of light used to view specimens. For example, shining polarized light onto a sample can reveal particular patterns of colour and structure in the same way that polarized sunglasses can. You can see this to great effect in the images of medicines in this book. Differential interference contrast uses two rays of polarized light, whose contrasting images of a sample are combined to reveal detail. This is particularly useful for studying transparent material.

Fluorescent light can also be used to show up unseen detail. Specific components within biological samples can be stained with fluorescent chemicals, which are visible in particular narrow wavelengths of light. The result is a fluorescence light micrograph. Immunofluorescence is the same technique applied specifically to antibodies in our immune systems. Multiphoton fluorescence miroscopes use longer wavelengths of light than usual – these use less energy and therefore cause less damage to cells under observation, an advantage when undertaking longer periods of observation or when studying living cells.

When viewing thick specimens through a microscope, especially at high magnification, it may not be possible to view the whole specimen in focus at once. There are various techniques for overcoming this problem. A confocal light micrograph is one in which out-of-focus parts of the specimen are omitted from the picture. It's an imaging technique particularly applied to fluorescence microscopy, where colourful but blurred background fluorescence might confuse matters. Deconvolution micrographs take a different approach; instead of omitting the blur, computer analysis works out what shape would have caused the blur and digitally 'restores' the whole picture in sharp focus.

Electron micrographs

At the start of the twentieth century scientists began to develop a high-tech alternative to the light microscope. The first electron microscopes appeared in the 1930s. Instead of beams of light, they use a stream of electrons fired from an electron gun. Instead of lenses they use electromagnets which can bend beams of electrons in the same way that glass lenses bend light. If the electron beam is dense enough, it becomes possible for the first time to see things in greater detail than with mere light – in other words, to see things that are not visible to the naked eye.

There are two kinds: the transmission electron microscope (TEM) and the scanning electron microscope (SEM). As its name suggests, the electrons from a TEM are transmitted – that is, they pass right through the material being studied. Because they pass through it, they are affected by it, just as light passing through stained glass is affected. It is the way in which the electrons are affected which builds up an image of the material, just as sunlight streaming through a stained-glass window lets us see the full colourful work of the window's designer. The TEM's image is collected on the far side of the material, either by camera or by a fluorescent screen.

By contrast the electrons from an SEM do not pass through the specimen. The SEM fires electrons which scan the specimen in a grid pattern. They interact with the atoms in the material, which then emit other electrons in response. These secondary electrons may be emitted in many directions depending on the shape and composition of the surface. They are detected; and by combining information from these secondary electrons with details of the original electron scan, a scanning electron micrograph is built up.

SEMs operate in a vacuum, although a variation, the environmental scanning electron microscope (ESEM), can see objects in gas or liquid – useful for viewing objects which would become unstable in a vacuum. Another variation, the ion-abrasion SEM, uses focussed beams of ions to peel away a thin section of material, like a tiny archaeological trench. This makes it possible to analyze, for example, the minute architectural structure within individual cells.

Because their electrons have to pass through the material, TEMs can only work with very thin samples of material. SEMs can deal with much bulkier material and the resulting images can convey depth of field. TEMs are however capable of greater resolution and magnification. The numbers are unimaginable, but TEMs can reveal details less than 50 picametres (50 trillionths of a metre/3¼ft) in width and magnify them over 50 million times. SEMs can 'see' details one nanometre (1000 picametres) in size and magnify them by up to half a million times. By comparison, an ordinary light microscope only shows detail larger than around 200 nanometres

4000 times larger than the TEM) and provides useful, undistorted magnification only up to two thousand times.

Microscopes are used to view samples of tissue removed from the human body. It is not always desirable, or possible, to separate tissue from the body; and in order to see things in situ within the body, different techniques are necessary.

X-ray radiographs

The commonest technique for looking inside the human body is x-ray radiography. X-rays were discovered in 1895 by German physicist Wilhelm Röntgen. Electromagnetic radiation bombards the body with x-rays, which are part of the electromagnetic spectrum, beyond ultraviolet light. The rays pass through the body and are collected as an image on the other side. Some of the x-rays are absorbed by denser parts of the body such as organs, and blocked altogether by bones. These parts appear as shadows in the image.

X-ray pictures are most commonly used to look at broken bones and foreign objects accidentally swallowed. Although they do not produce highly detailed pictures, simple x-rays also help to diagnose illnesses. For example the respiratory tract can appear more opaque in cases of tuberculosis and pneumonia.

Sometimes patients are asked to drink a ‘meal’ of barium sulphate before an x-ray. Its presence in the digestive system will show up under x-rays, revealing any abnormalities in the gut. A similar process produces images called angiograms. In this case so-called contrast agents are injected into the blood supply. Like barium, these agents work by blocking x-rays, and angiograms show the arteries, veins and other blood vessels as dark

well-defined networks in which irregularities such as constriction or blockage can be identified.

Scintigraphy is, in a sense, the opposite of x-ray radiography. Scintigrams are produced by consuming radioactive isotopes. The isotopes emit radiation from within the body, unlike x-rays which bombard it from without. Cameras around the body detect the emissions, which reveal the pathways travelled by the isotopes. A concentration of radiation may indicate a blocked or constricted pathway.

CT and MRI scans

In the 1970s computer technology was applied to x-ray images in a process called computed tomography (CT). CT scans use x-rays from several directions and compile the results to create a computer-generated cross-section of the body. The image, made with information scanned from several different viewpoints, is significantly more detailed than a simple x-ray radiograph.

There is a moderate risk of damage from the higher doses of x-rays involved in CT scans. Magnetic resonance imaging (MRI) scans, available since the 1980s, carry no such risks, although they are not suitable for patients with certain implants. MRI scanners use powerful electromagnets to stimulate the hydrogen molecules in water within the body. By detecting the radio frequency which the molecules emit, and the speed with which they settle down again, the scanner can build a detailed map of the body through a series of cross-sections.

CT and MRI scans enable us to search for abnormalities such as tumours without invasive surgery, a clear advantage for both patients and doctors. MRI techniques such as diffusion tensor imaging and diffusion spectrum imaging go further, tracing the movement of water molecules, particularly in the nervous system. This has brought us extraordinary new understanding of how our brains work by producing pictures of the neural pathways of our minds.

Colour

Of course if you were to examine the human body yourself, it wouldn't look like most of the strikingly colourful pictures in this collection. Inside us all, we are mostly just various shades of flesh and blood. Indeed, many of the body's cellular structures are transparent and have no colour at all. Many of the images here have therefore been artificially coloured to help identify the various elements in them. While experts may understand the pinkish mass of membranes and cavities in a section of the lung, for example, a little colour-coding for the rest of us is very helpful.

Colouring like this is easily added to an image by computer. But sometimes colour is applied to the biological samples themselves. As well as the fluorescent dyes used to produce fluorescence light micrographs, biologists introduce a wide range of stains to highlight individual components and make them visible for the camera. Particular stains can be targeted to reveal specific proteins, for example, either by sticking only to the protein in question (called positive staining) or by sticking to everything except the protein (negative staining). Negative staining is preferable, because at the microscopic level even a tiny amount of positive stain may obscure some detail. Either way, colour is a useful tool for helping both the biologist and the public at large to understand the minute and complex systems of the human body.

We can now see the very smallest working parts of our bodies. Our ability to do so is driven by a desire to understand how they work, and how to fix them when they don't. We live our lives in a remarkable machine, one which can repair and regenerate its own parts, which moves both deliberately and instinctively, with and without conscious thought; which learns how to do things better; which can defend itself from viruses and bacteria and from almost anything we can throw at its digestive system; which reacts not only to the physical environment in which it lives but to sources of pleasure such as art and love which have no direct bearing on its ability to function. The human body is an astonishing, complicated, wonderful home. And as this book proves, its science is beautiful.

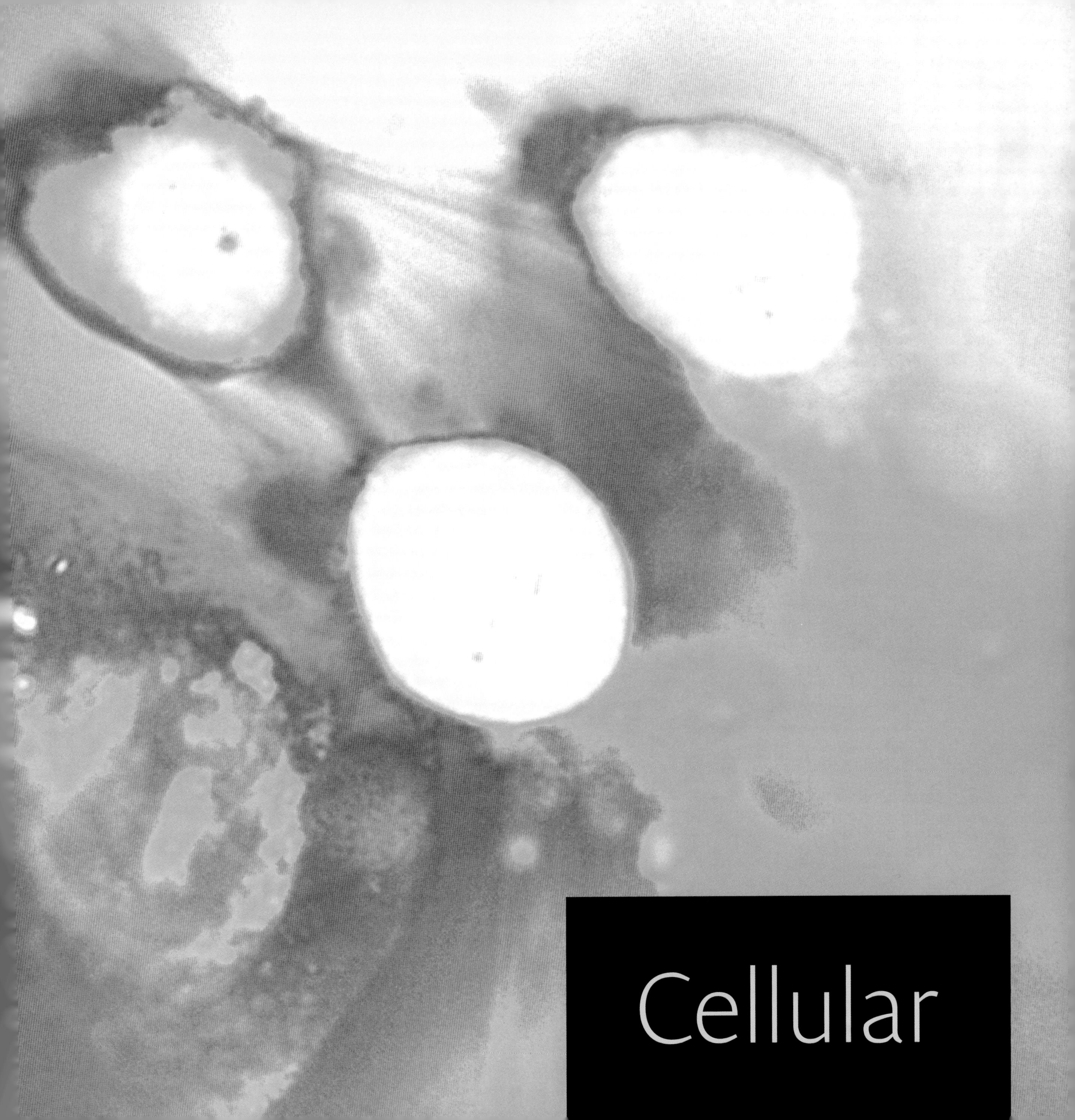
Cellular

Previous page: Human nerve cells in culture (light micrograph)

It is common practise to grow nerve cells in controlled laboratory conditions, in order to study the workings of the central nervous system. Networks of such neurons can be connected to a computer via electrodes which allow researchers to send and receive signals from the cells. This beautiful image of such cells was achieved by optical staining – a way of 'seeing' objects by measuring their appearance under different wavelengths of light. (Magnification: unknown)

Sperm production (coloured scanning electron micrograph)

Sperm is produced in a number of highly folded chambers called seminiferous tubules within each testicle. The tubule in this image contains a swirl of the tails (blue) of developing sperm cells at its centre. The lining of the tubule contains two types of cell: Sertoli cells (red), which nourish developing sperm, and spermatogenic cells (green), from which the sperm cells form. This view was achieved by freeze-fracture – the rapid freezing and breaking open of fresh tissue. (Magnification: unknown)

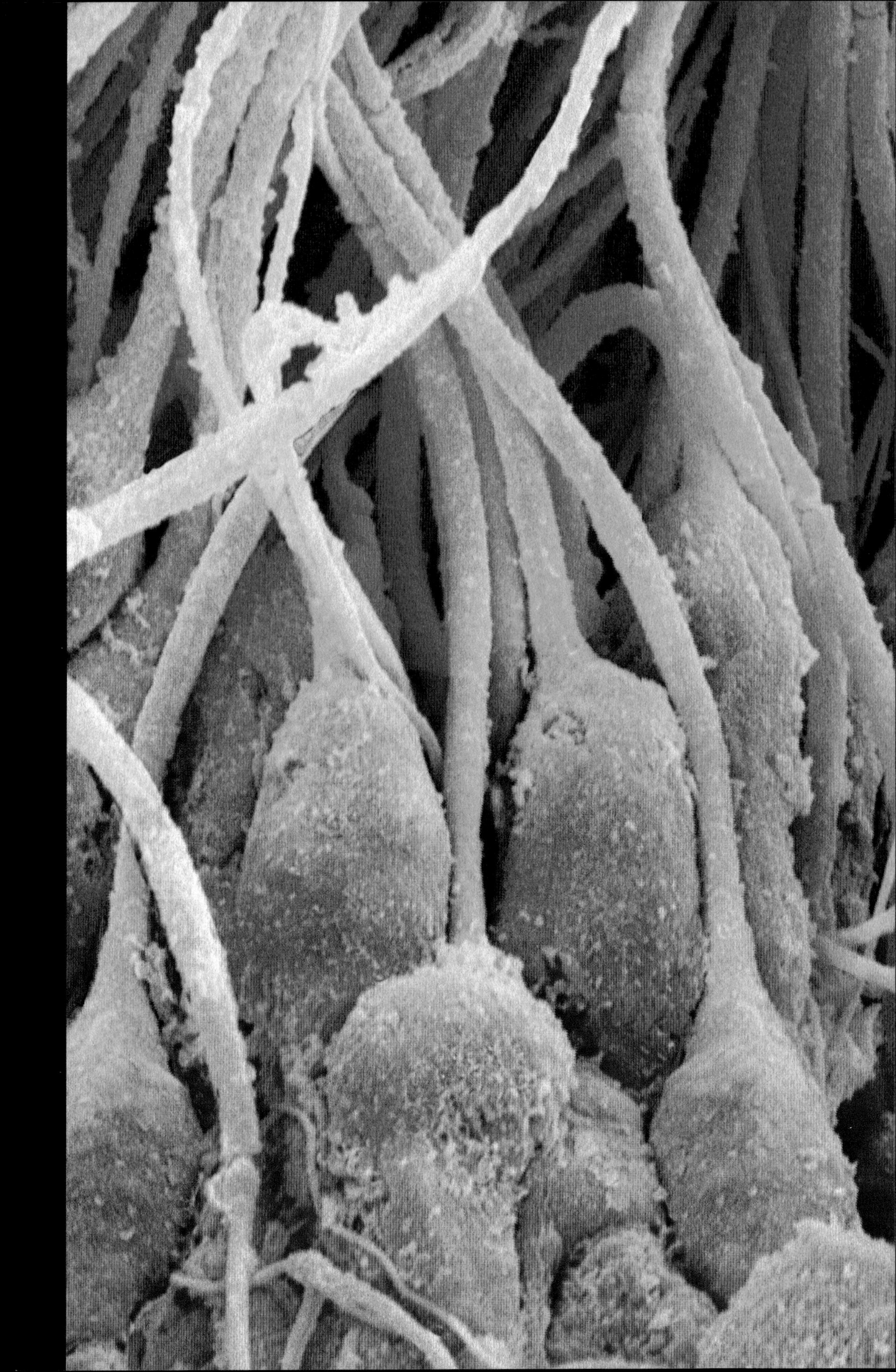

Sperm production (coloured scanning electron micrograph)

Sperm cells (blue tails) develop inside a seminiferous tubule, the site of sperm production in the human testis, in a process called spermatogenesis. The developing heads of the sperm cells (green/red, across bottom) are embedded in a layer of Sertoli cells (red) which provide nutrients for the developing sperm. It is the heads that contain the genetic material used to fertilize the female egg cell. The process by which sperm mature into their mobile form is called spermiogenesis. (Magnification: x3750 at 10cm/4in size)

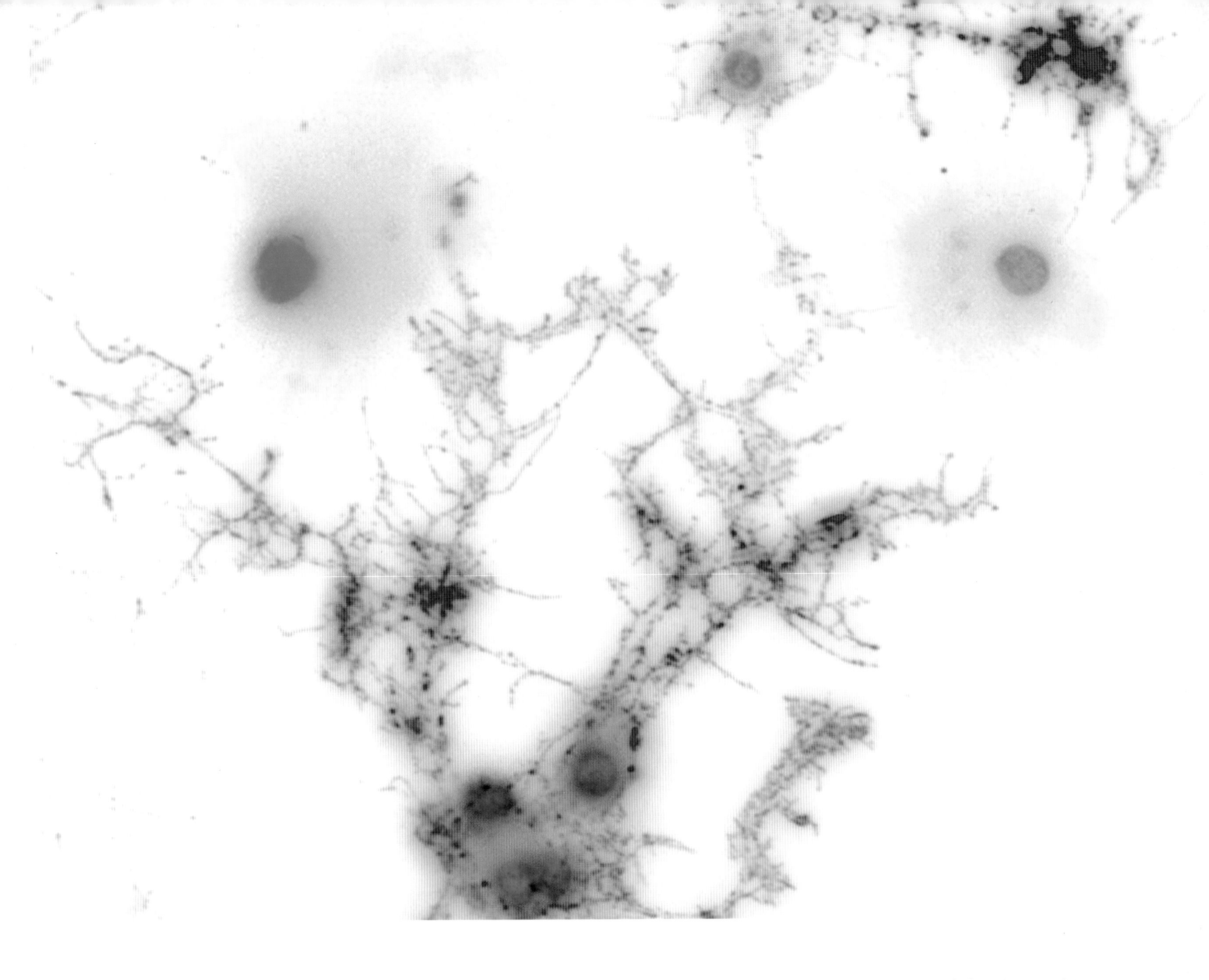

Above: Brain cells (fluorescence light micrograph)
Less than half the cells of the central nervous system in the brain are neurons (nerve cells). The rest have supporting roles, of which these are two examples. White blood cells known as phagocytic macrophages (here stained yellow) detect microbes and other threats to the system, then swallow and dispose of them. The main function of oligodendrites (here stained red) is to insulate the axon, the part of a neuron which transmits information in electrical impulses. (Magnification: x40 at 10cm/4in size)

Right: Nerve cells in the brain (fluorescence light micrograph)
Nerve cells occur in the brain and spinal cord (the central nervous system), and in ganglia – clusters of nerve cells outside the system. Each nerve cell has a large cell body (here shown as orange) with several long extensions (green), called processes because they process information. They usually consist of several thinner branched dendrites (which receive information from other nerve cells) and one thicker axon (which passes it on after the cell body has interpreted it). (Magnification: unknown)

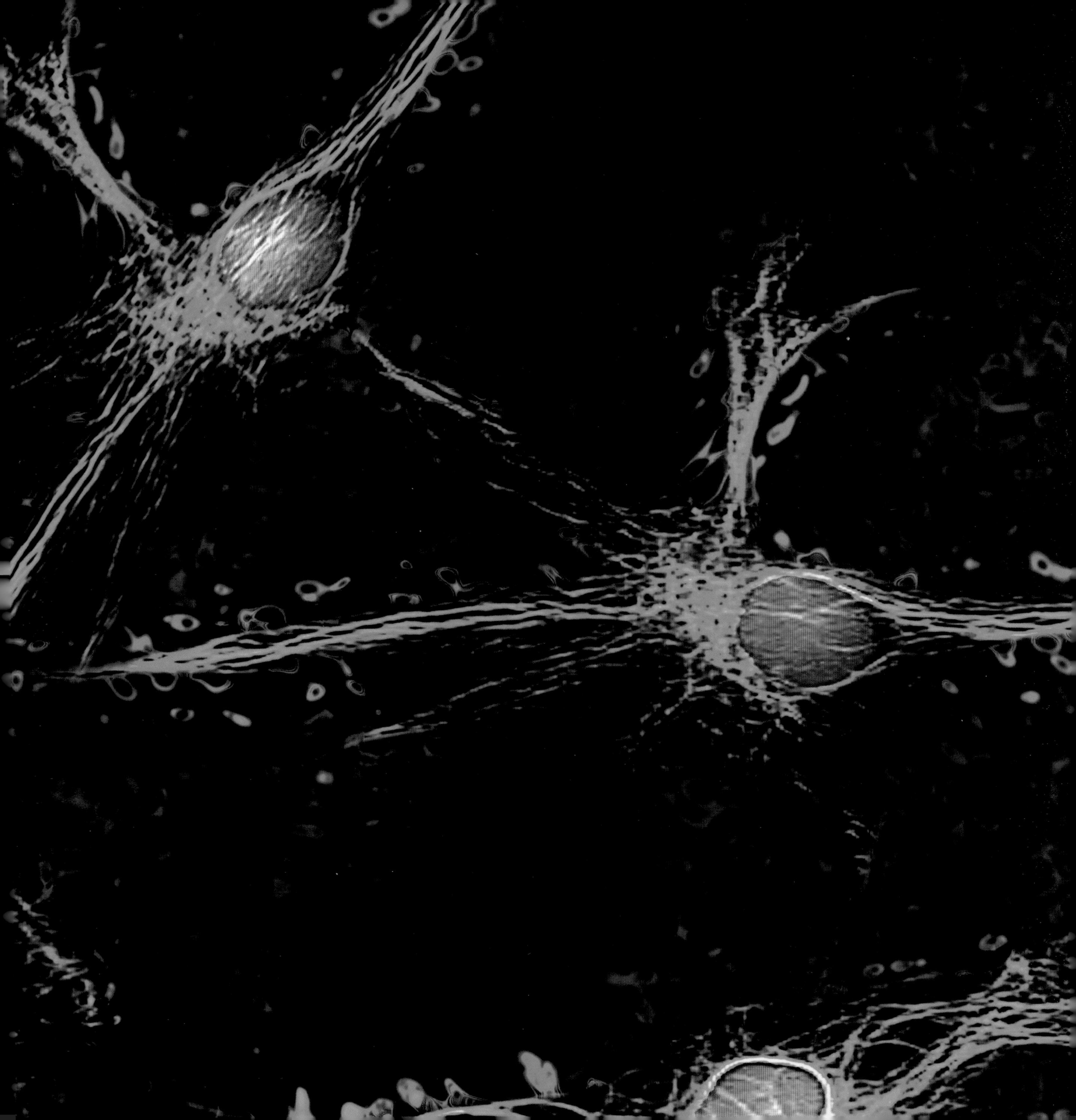

Astrocyte brain cells (fluorescence light micrograph)
Astrocytes have limb-like branches that provide support and nutrition to the neurons (nerve cells) of the brain. In this image, of the brain of a human foetus, the cell nuclei are lilac; the green fronds are their protein tendrils, through which dendrites and axons (a neuron's receptors and transmitters) pass as they grow, gathering nutrients. Astrocytes also play a remedial role, repairing damage to neurons from microbes, toxins or loss of blood. (Magnification: x40 at 10cm/4in size)

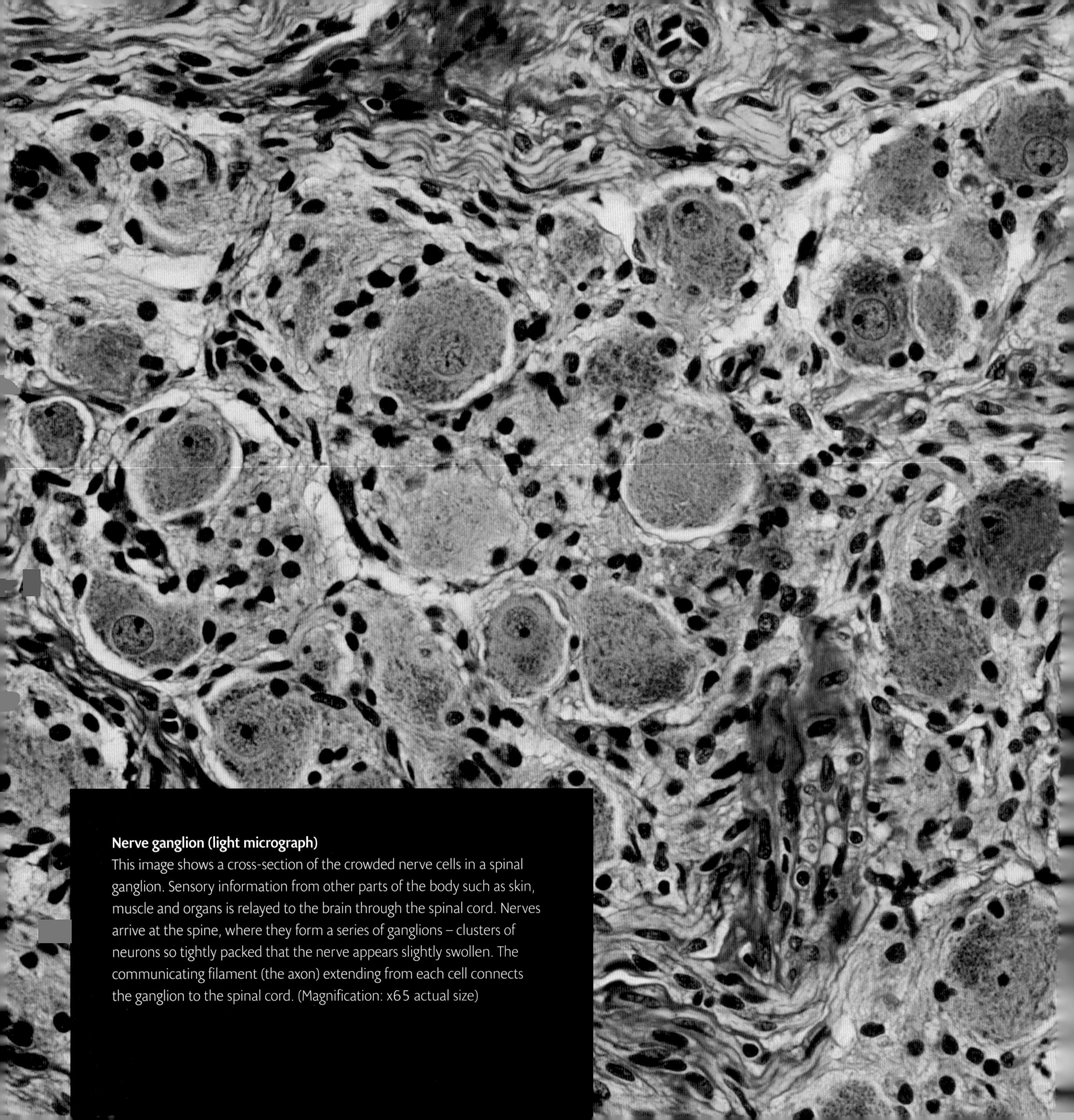

Nerve ganglion (light micrograph)
This image shows a cross-section of the crowded nerve cells in a spinal ganglion. Sensory information from other parts of the body such as skin, muscle and organs is relayed to the brain through the spinal cord. Nerves arrive at the spine, where they form a series of ganglions – clusters of neurons so tightly packed that the nerve appears slightly swollen. The communicating filament (the axon) extending from each cell connects the ganglion to the spinal cord. (Magnification: x65 actual size)

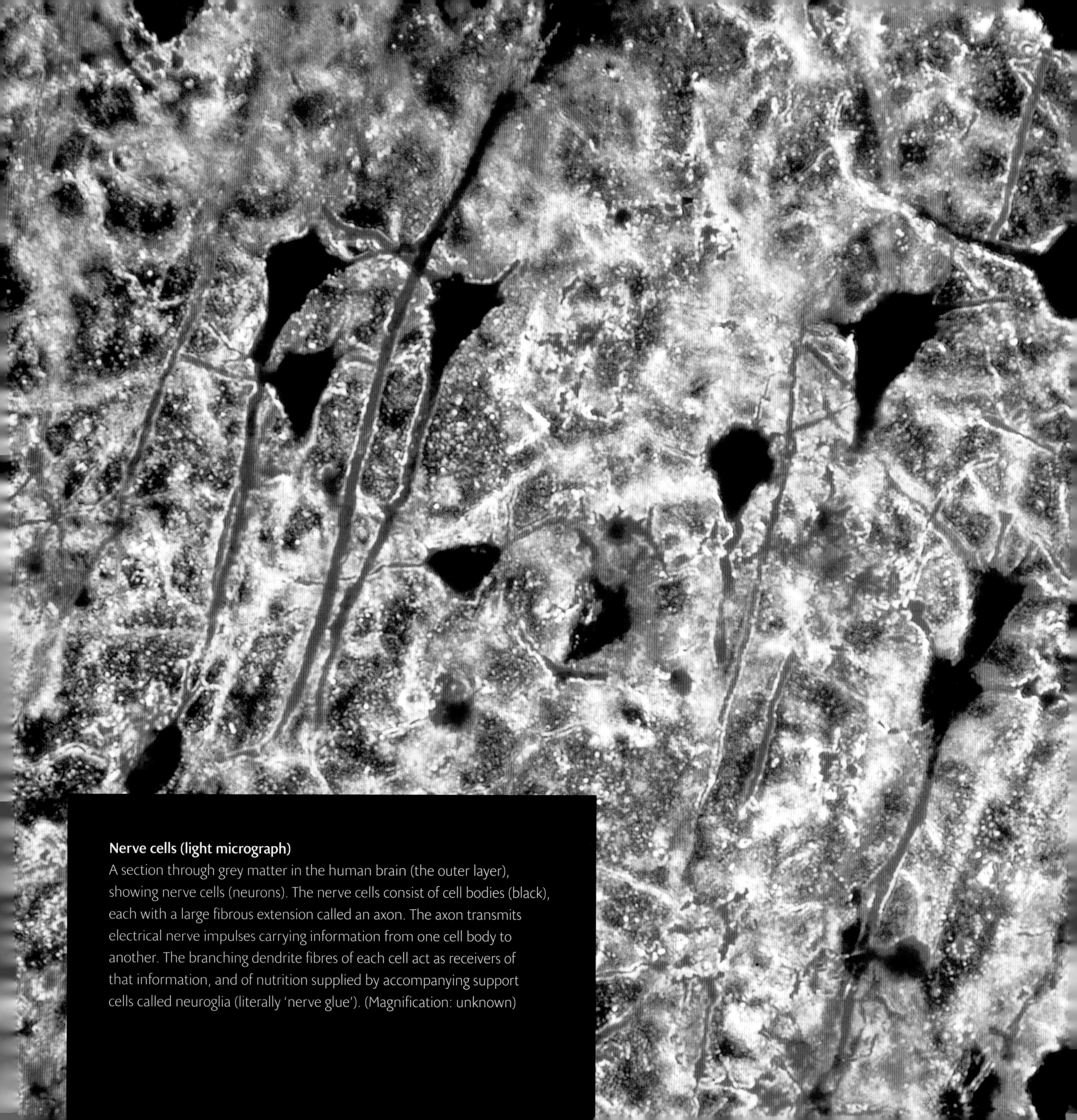

Nerve cells (light micrograph)
A section through grey matter in the human brain (the outer layer), showing nerve cells (neurons). The nerve cells consist of cell bodies (black), each with a large fibrous extension called an axon. The axon transmits electrical nerve impulses carrying information from one cell body to another. The branching dendrite fibres of each cell act as receivers of that information, and of nutrition supplied by accompanying support cells called neuroglia (literally 'nerve glue'). (Magnification: unknown)

Nerve cell growth (fluorescence light micrograph)

Nerve cells grown in the laboratory show healthy signs of growth here, stimulated by a protein called nerve growth factor. The yellow tendrils with blue branches are the neurites, through which each cell communicates with other cells. Each cell body contains a nucleus (pink). Laboratory research into how nerves regenerate is helping us to understand and cure spinal paralysis. These examples are PC12 cells, from a tumour of the adrenal gland, which are able to sense calcium concentrations. (Magnification: x670 at 10cm/4in size)

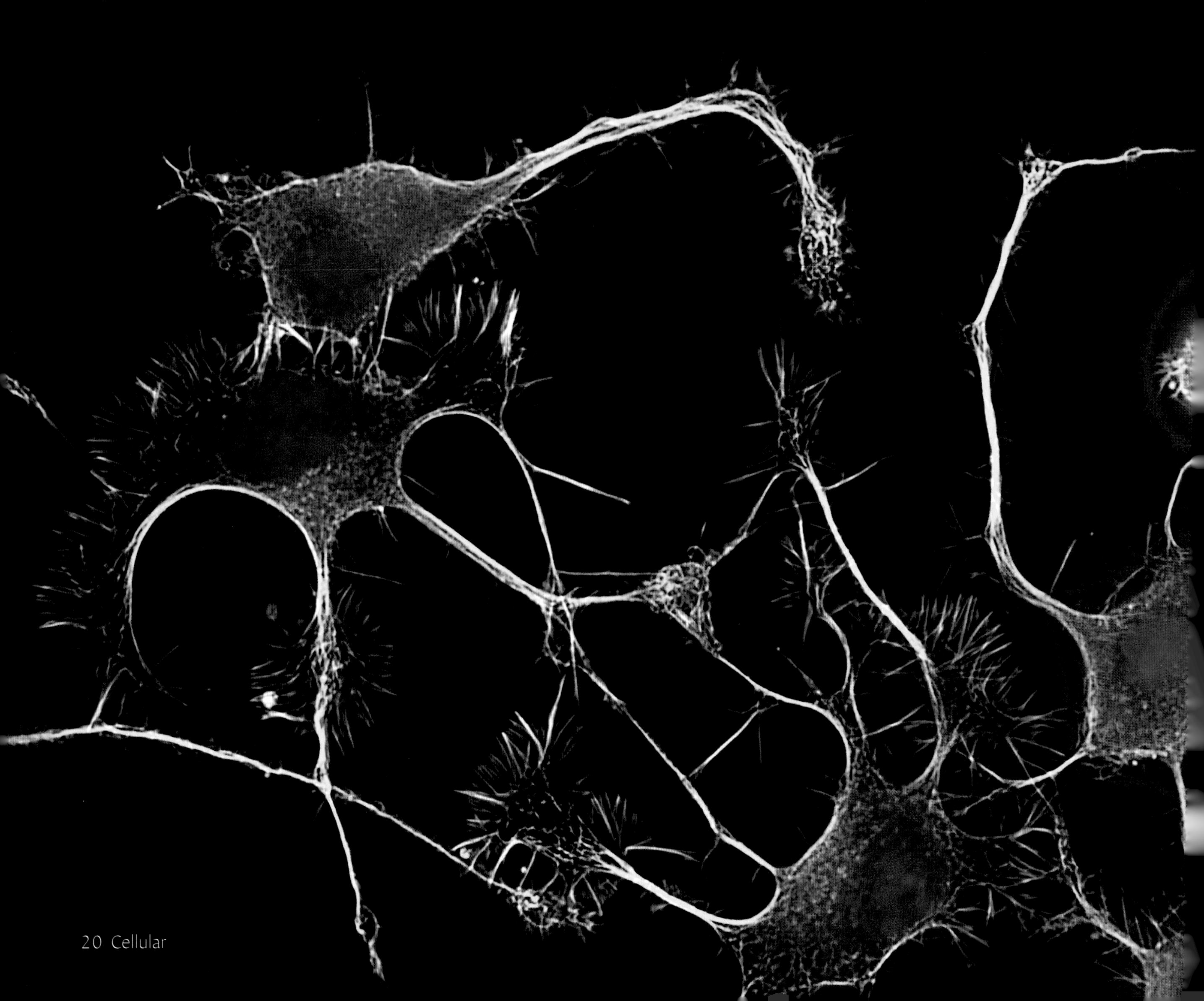

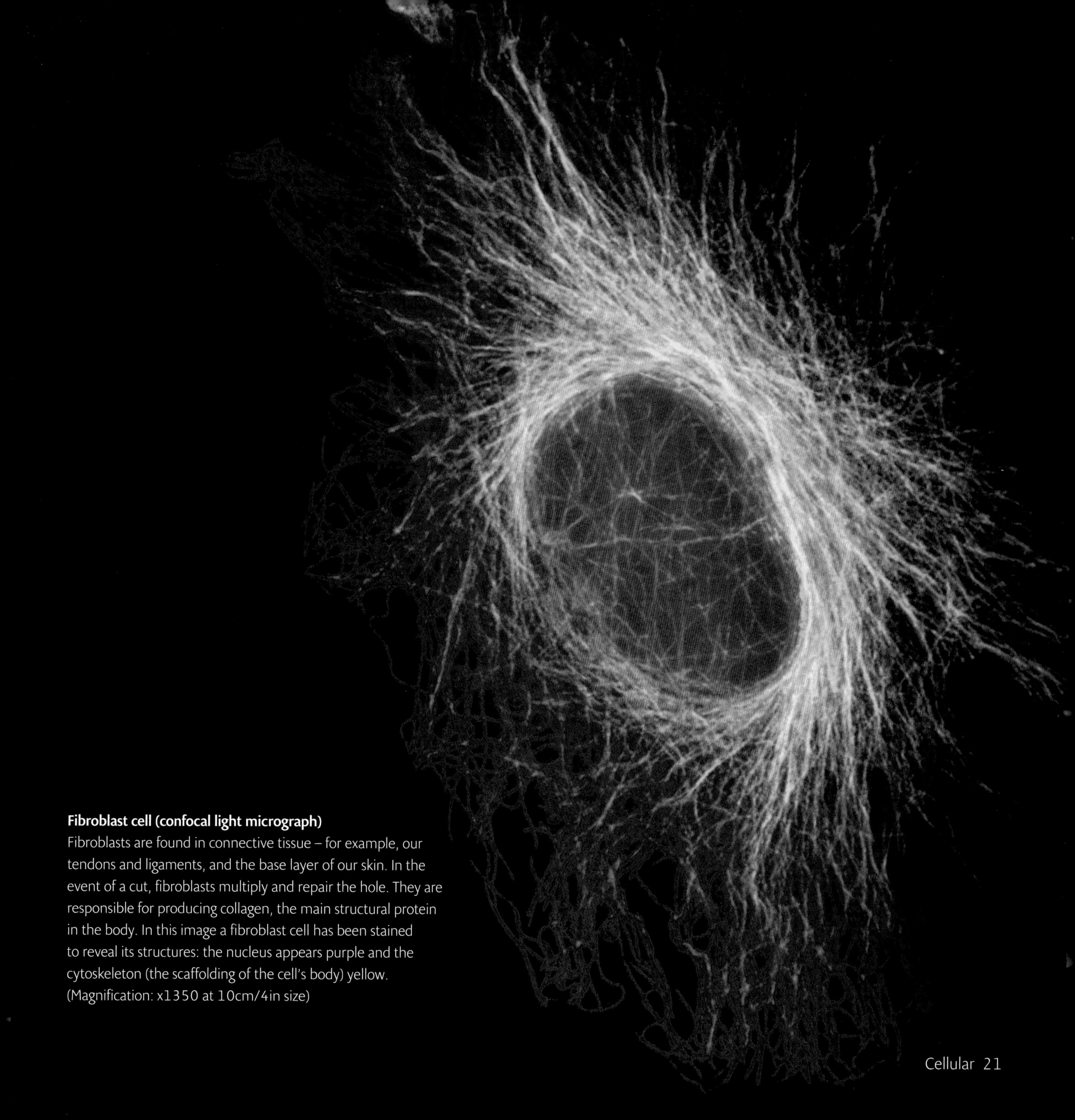

Fibroblast cell (confocal light micrograph)
Fibroblasts are found in connective tissue – for example, our tendons and ligaments, and the base layer of our skin. In the event of a cut, fibroblasts multiply and repair the hole. They are responsible for producing collagen, the main structural protein in the body. In this image a fibroblast cell has been stained to reveal its structures: the nucleus appears purple and the cytoskeleton (the scaffolding of the cell's body) yellow. (Magnification: x1350 at 10cm/4in size)

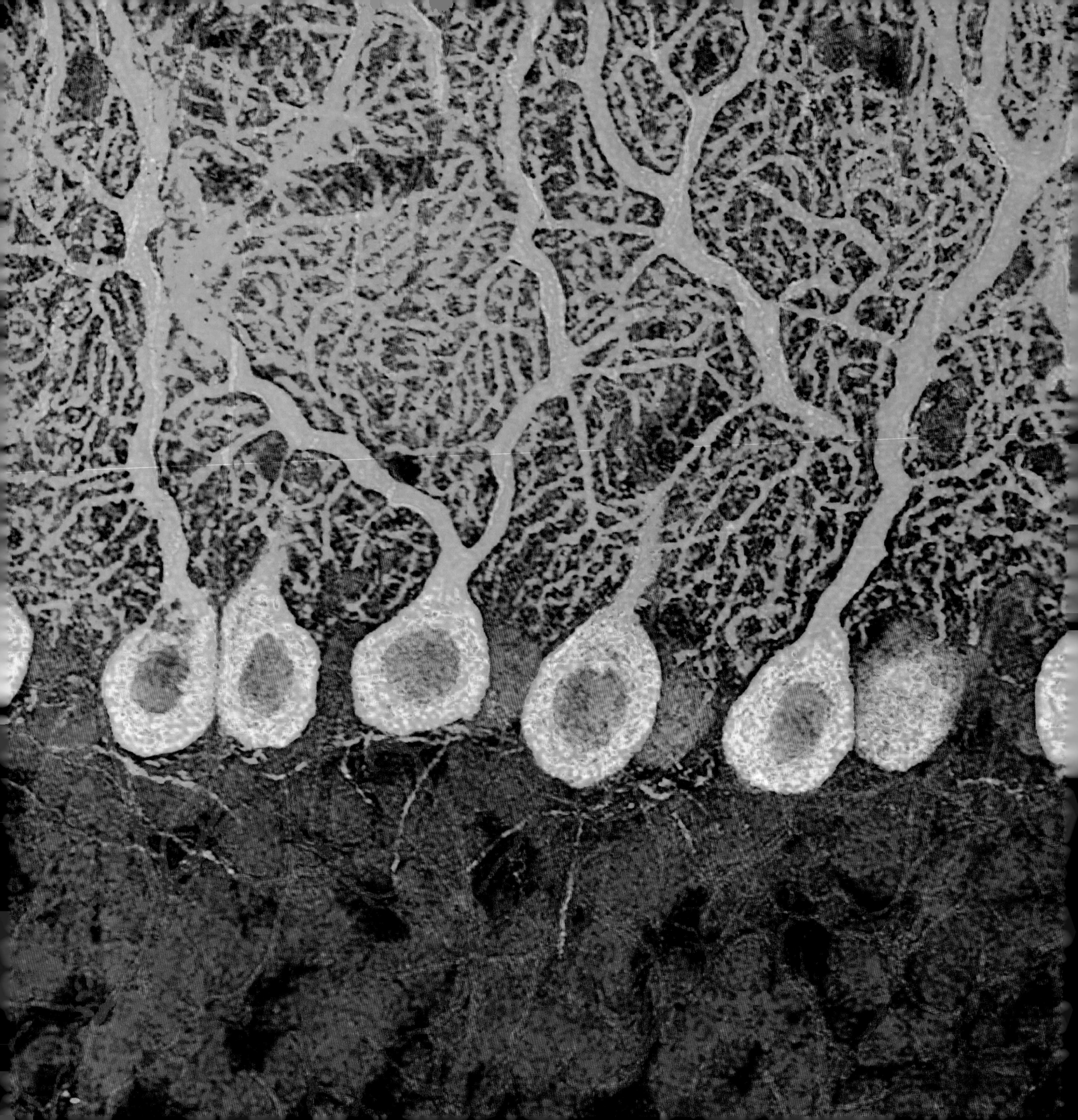

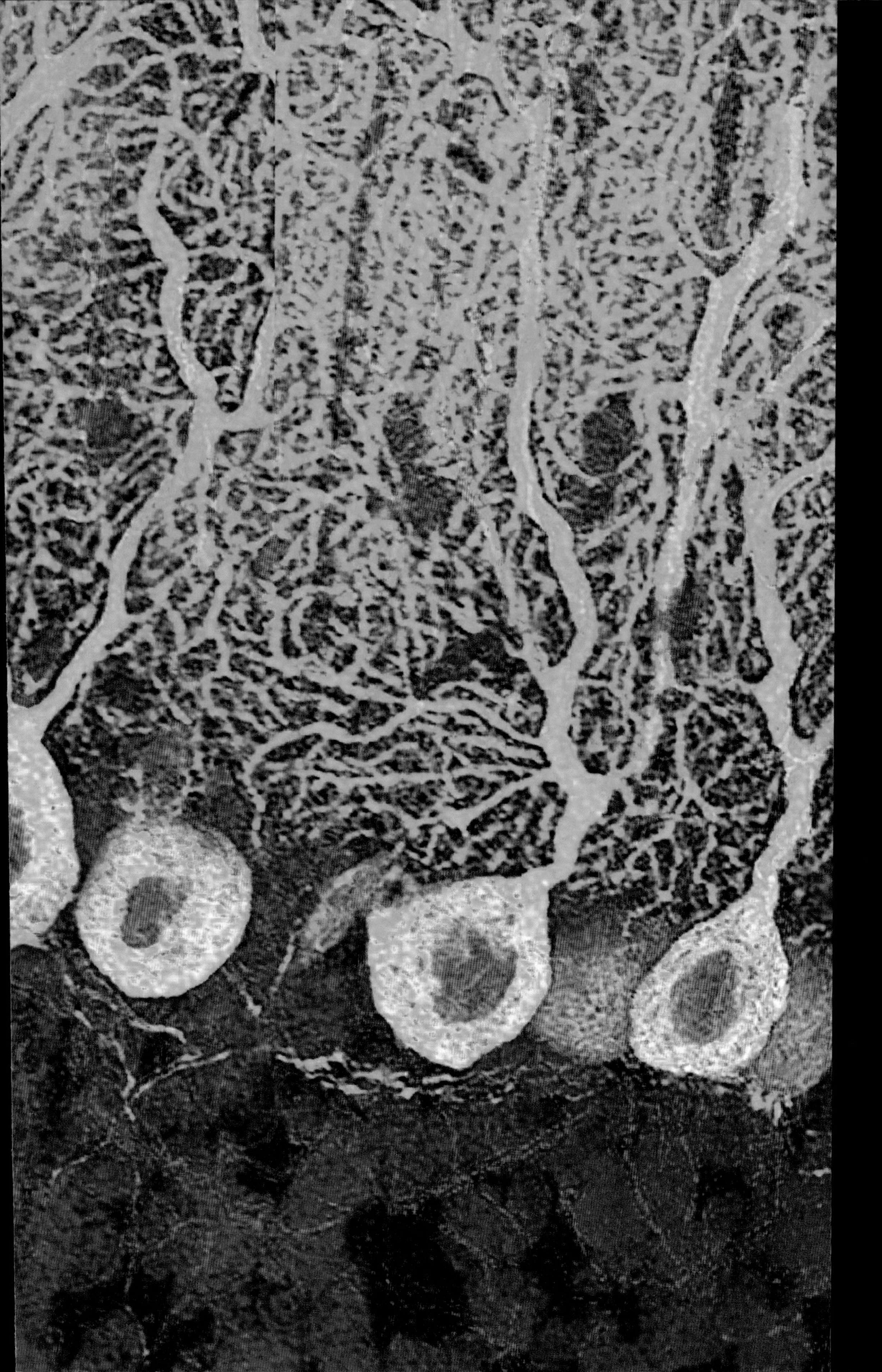

Purkinje cells in the cerebellum (fluorescence light micrograph)

These characteristic coral-like structures are Purkinje nerve cells in the cerebellum of the brain. They are found in the grey matter (cortex) of the cerebellum, at the boundary between two layers. Purkinje cells have a flask-like body (yellow, with a red nucleus) from which numerous highly branched information-gathering dendrites (green) extend upwards into the outer layer (red green), and a single axon (here visible as a fine green line) relays information to the inner layer (blue red). (Magnification: x380 at 10cm/4in size)

Nerve cell culture (coloured scanning electron micrograph)

A nerve cell shares sensory information with other nerve cells through its neurites – the branching extensions seen here which send or receive that information. These neurites have grown from a cultured sample of cells (not visible in this image) from a spinal cord, as part of research into curing spinal paralysis. The research uses nerve tissue cultures like this to investigate ways of encouraging neurite growth and nerve regeneration. (Magnification: unknown)

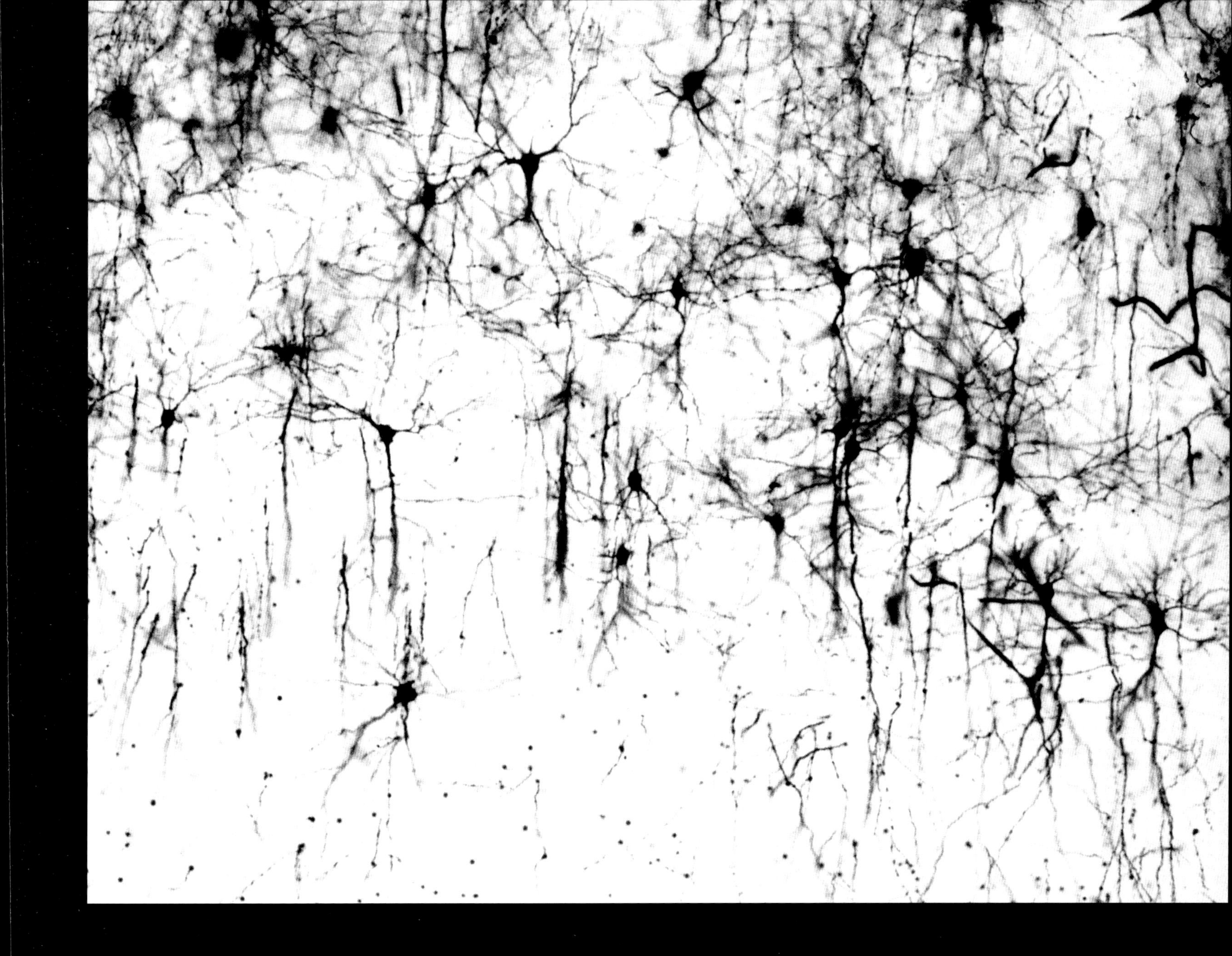

Cerebral cortex nerve cells (light micrograph)

The grey matter of the cerebral cortex forms the outer layer of the brain that is responsible for conscious thought, memory and language. These cobweb-like structures within the cortex are neurons. The dark spots are the bodies of the nerve cells, from which their receptors (branching dendrites) and transmitters (single axons) extend in a tangle of fine lines.

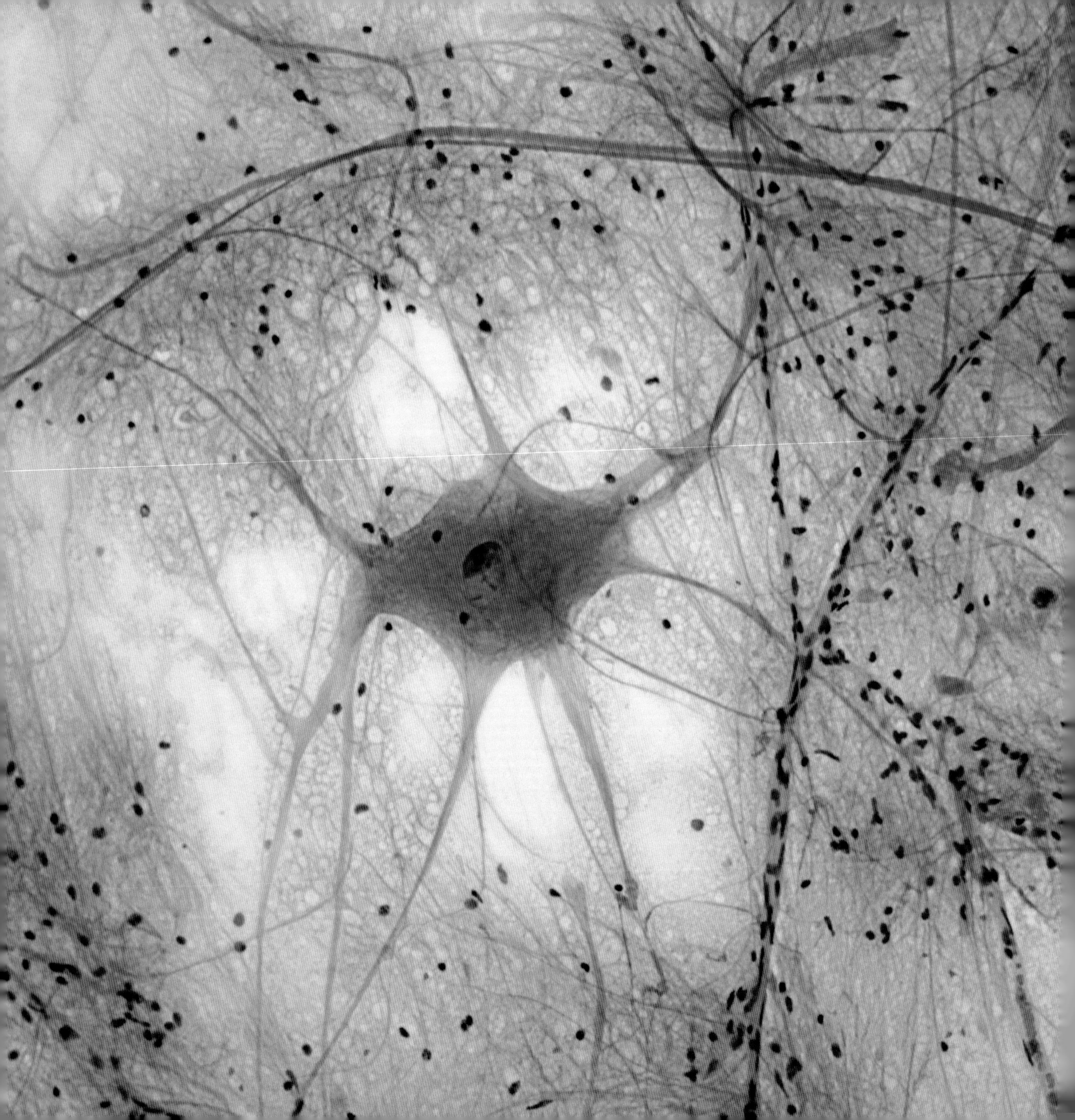

Left: Multipolar neurons (light micrograph)
A cluster of multipolar nerve cells (neurons) from the grey matter of a human spinal cord. The branching 'arms' leading from the cell body in the centre are dendrites, receiving impulses from other cells. Cells are classified as multipolar when, as here, many dendrites stem directly from the body. By contrast a bipolar cell only has two extensions – one dendrite and one axon; and a unipolar cell only has one, from which both the axon and the dendrites then branch. (Magnification: x280 at 10cm/4in size)

Above: Dendritic cell (coloured ion-abrasion scanning electron micrograph)
Dendritic cells, looking like blousy peony blossoms, are a component of the body's immune system. Their sheet-like extensions maximize their surface area, trapping foreign molecules and presenting them to other cells in the immune system to be eliminated. HIV (human immunodeficiency virus) uses the body's defences against itself by taking advantage of this procedure. When dendritic cells carrying HIV cells interact with T cells (white blood cells matured in the thymus gland), HIV is transmitted to the T cells, infecting them. (Magnification: unknown)

Above: Neural progenitor cell differentiation (fluorescence light micrograph)

Neural progenitor cells are able to develop into either neurons (nerve cells) or support cells (glial cells) such as astrocytes, the cells which provide nutrition for neurons. In this image they have been grown for three weeks in an environment that favours astrocytes. The progenitor cells appear green, the astrocytes yellow. The blue cores in both are their nuclei. Neural progenitor cells are a potential source of cells to replace damaged or lost brain cells. (Magnification: unknown)

Right: Hormone receptor nerve cells (confocal light micrograph)

The small round lilac cells in this image are receptors in the brain for the orexin hormones. Their nuclei appear pink, as do the much larger nuclei of neighbouring green nerve cells. Orexins, also known as hypocretins, are hormones that regulate the sleep cycle by inducing wakefulness. They also play a role in stimulating appetite. The two names are the result of their simultaneous discovery by two groups of researchers, who have yet to agree which one to use. (Magnification: unknown)

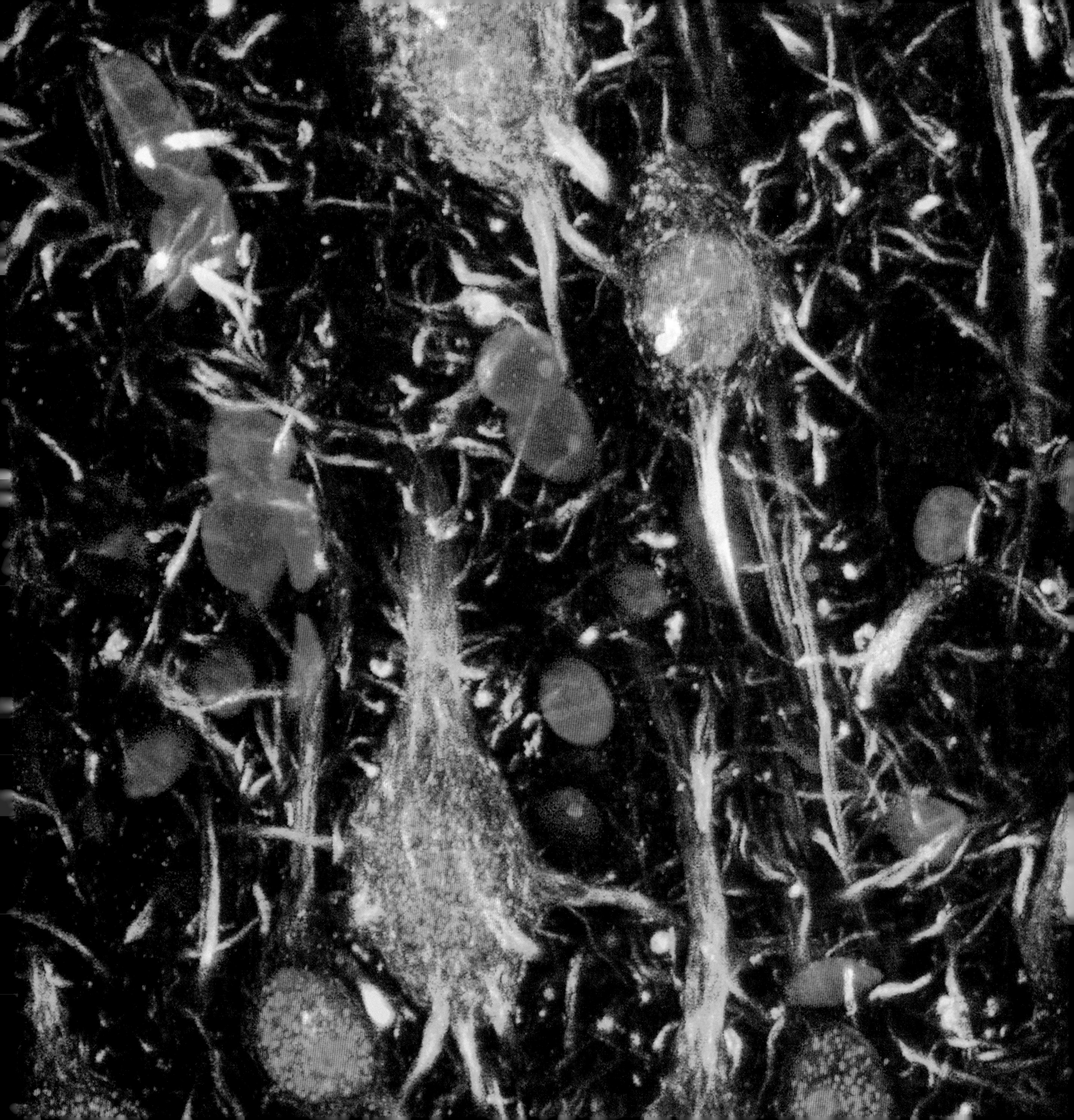

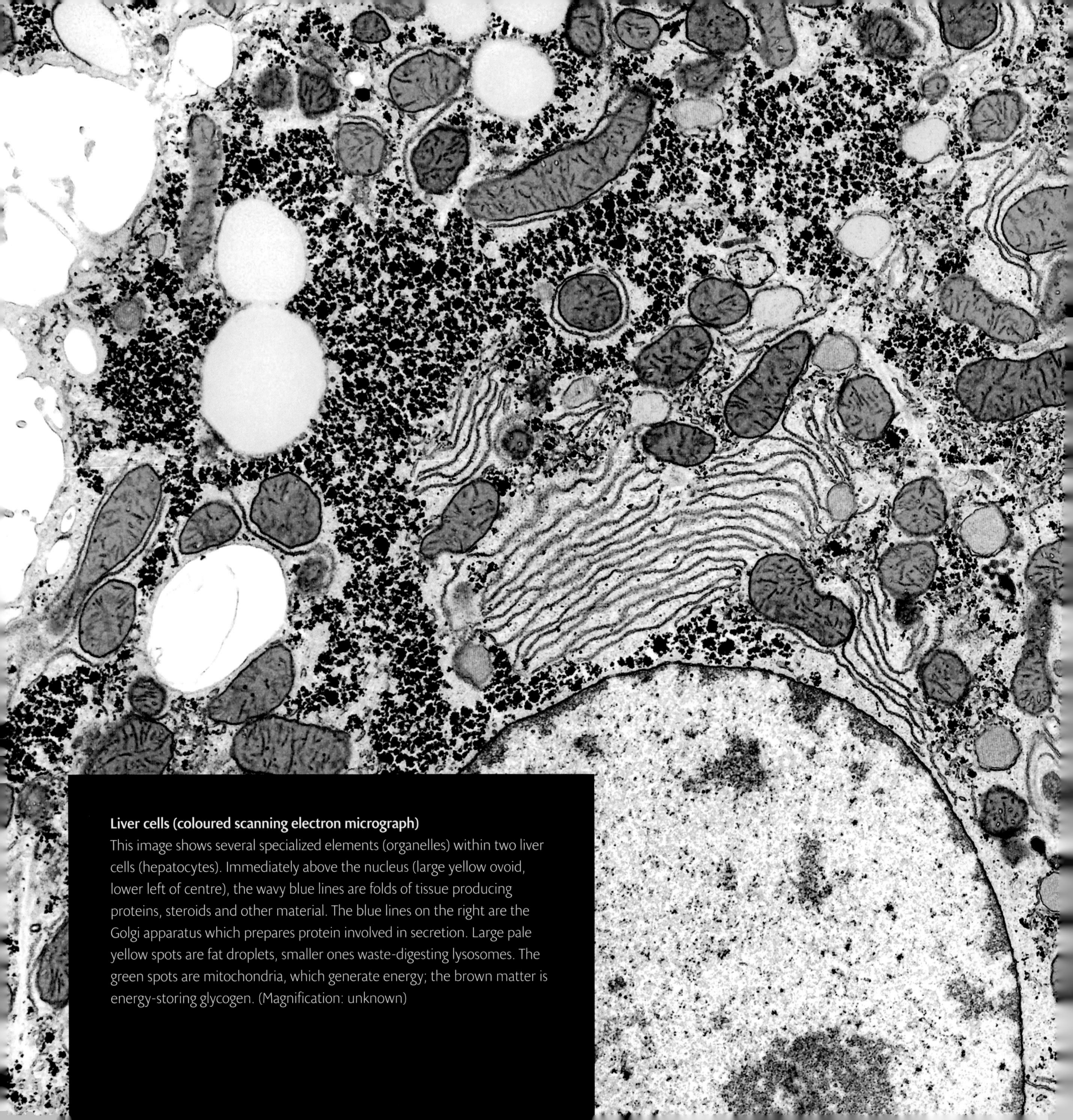

Liver cells (coloured scanning electron micrograph)
This image shows several specialized elements (organelles) within two liver cells (hepatocytes). Immediately above the nucleus (large yellow ovoid, lower left of centre), the wavy blue lines are folds of tissue producing proteins, steroids and other material. The blue lines on the right are the Golgi apparatus which prepares protein involved in secretion. Large pale yellow spots are fat droplets, smaller ones waste-digesting lysosomes. The green spots are mitochondria, which generate energy; the brown matter is energy-storing glycogen. (Magnification: unknown)

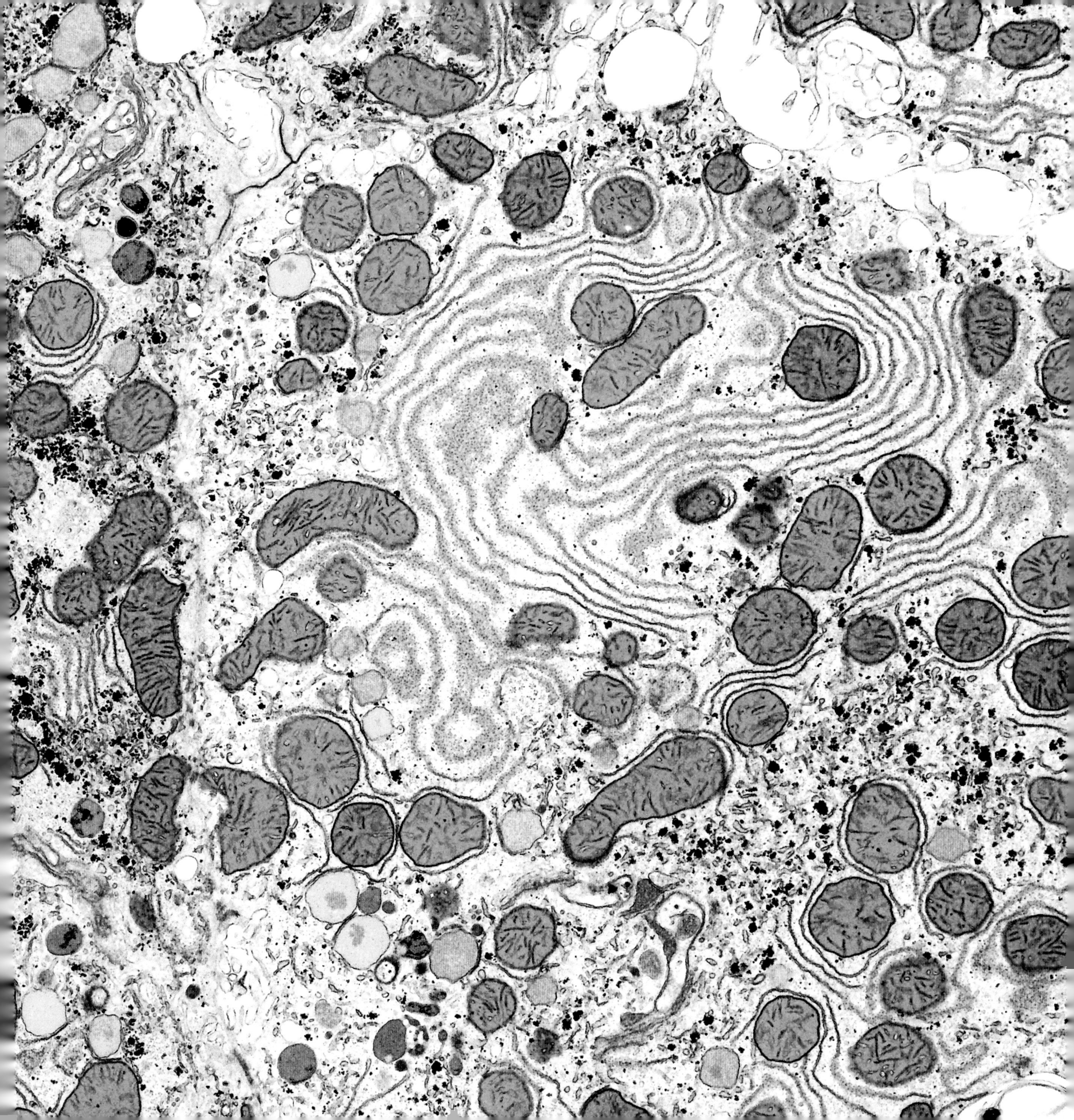

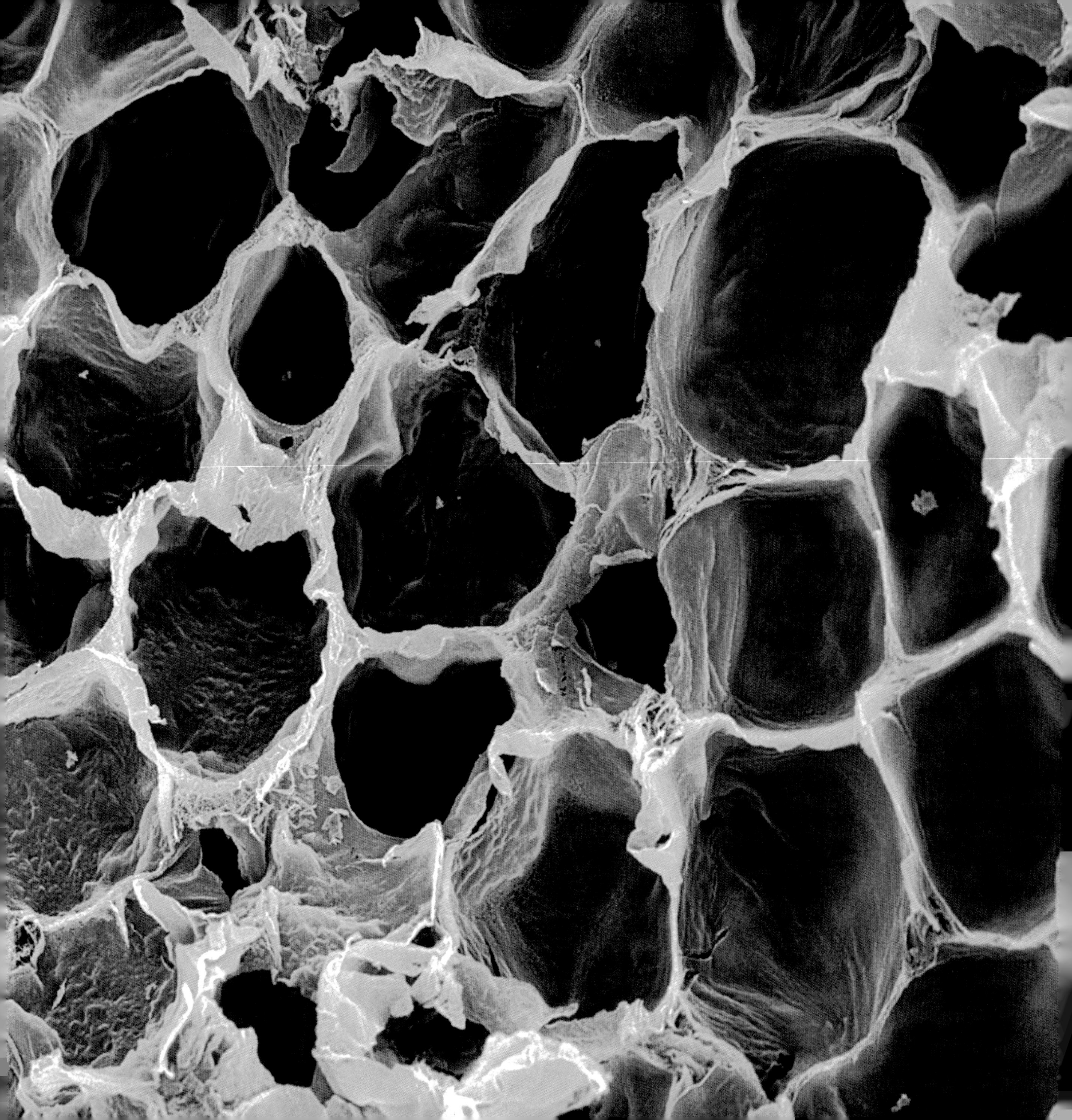

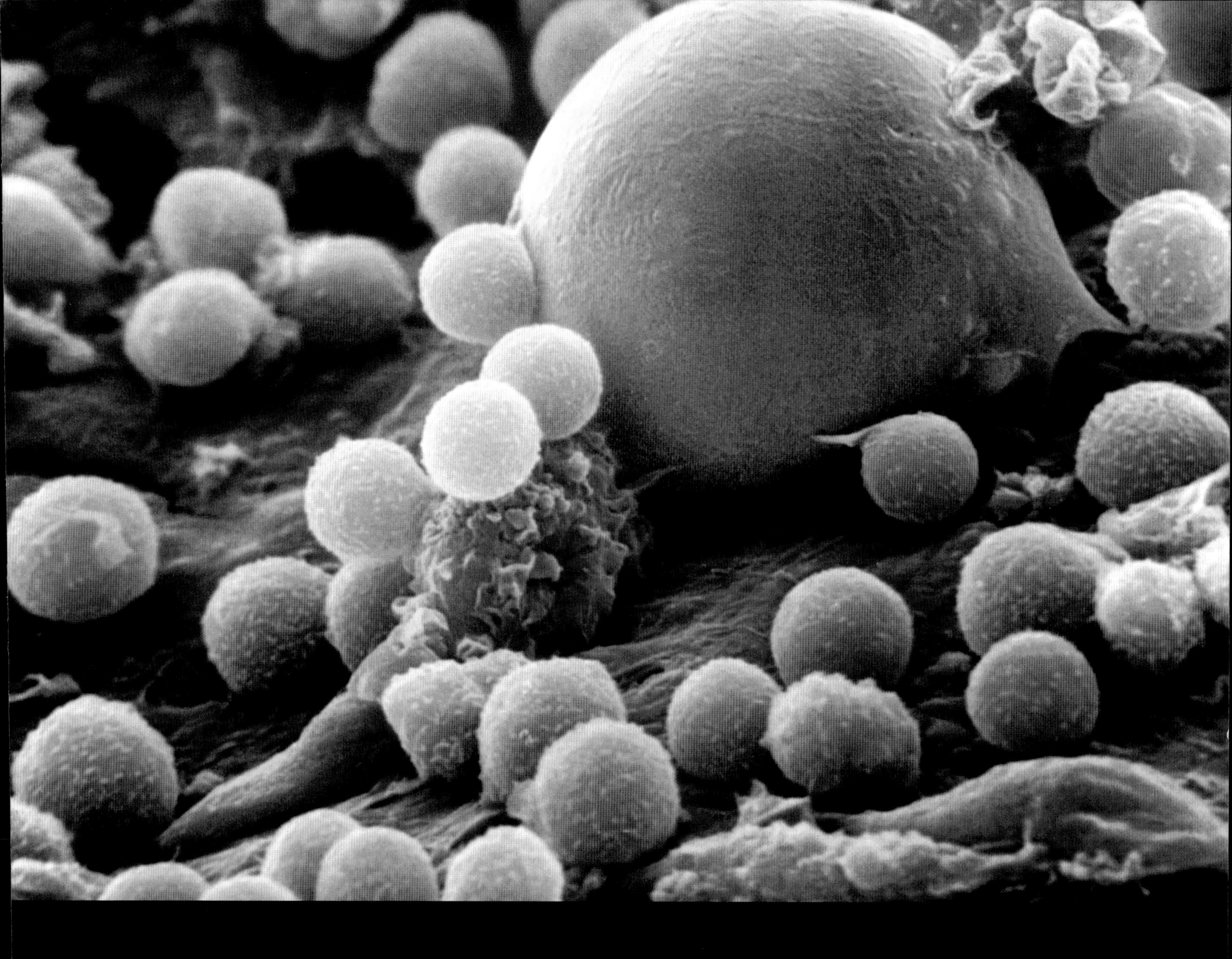

Left: Empty fat cells (coloured scanning electron micrograph)
Fat cells (adipocytes) are amongst the largest cells in the human body. They form a thick insulating layer under the skin which serves to cushion us as well as to store energy. In this image the normal lipid (fat) deposits of the cells (their major component) have been removed, revealing the honeycomb structure of the cell membranes. When we put on weight, the cells swell with additional fat, and eventually extra cells are added too. (Magnification: x350 at 6x7cm/2¼x2¾in size)

Above: Fat cell (coloured scanning electron micrograph)
Fat cells (adipocytes) store energy as an insulating layer of fat and the majority of the cell's volume is taken up by a large lipid droplet. Here, an adipocyte (blue) from bone marrow tissue is surrounded by two types of white blood cell: granulocytes (purple and round) and macrophages (green and frilly). Both are phagocytes – cells which fight infection by engulfing alien organisms and breaking them down into harmless waste. (Magnification: unknown)

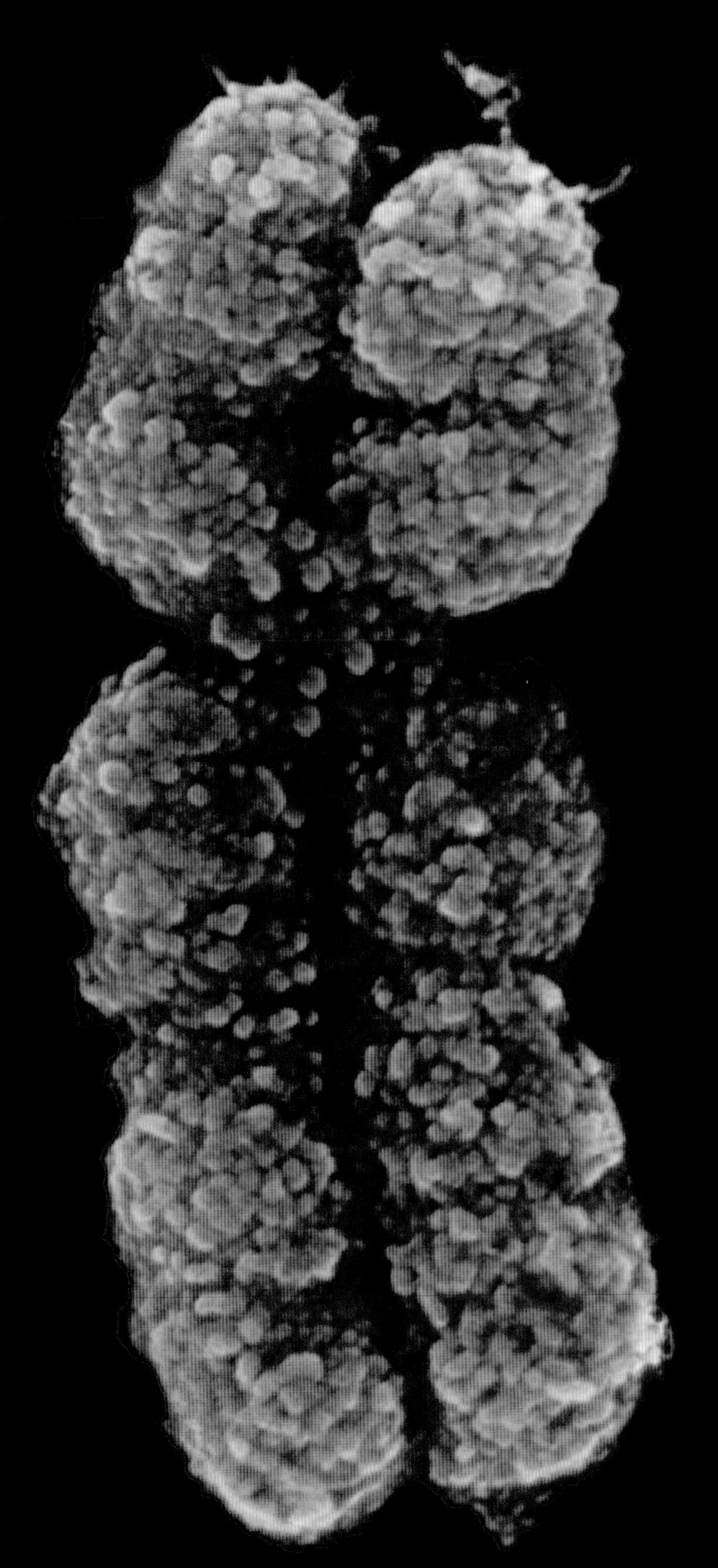

X and Y chromosomes (coloured scanning electron micrograph)
Humans have 46 chromosomes: 23 inherited from the mother and 23 from the father. They are paired, and each pair contains particular pieces of genetic information – for example pair 11 is associated with our sense of smell, but also with autism and our resistance to breast cancer. Pair 23 consists of the sex chromosomes which determine an individual's gender. Males have an X sex chromosome (pink) and a Y sex chromosome (blue) and females have two X sex chromosomes. (Magnification: unknown)

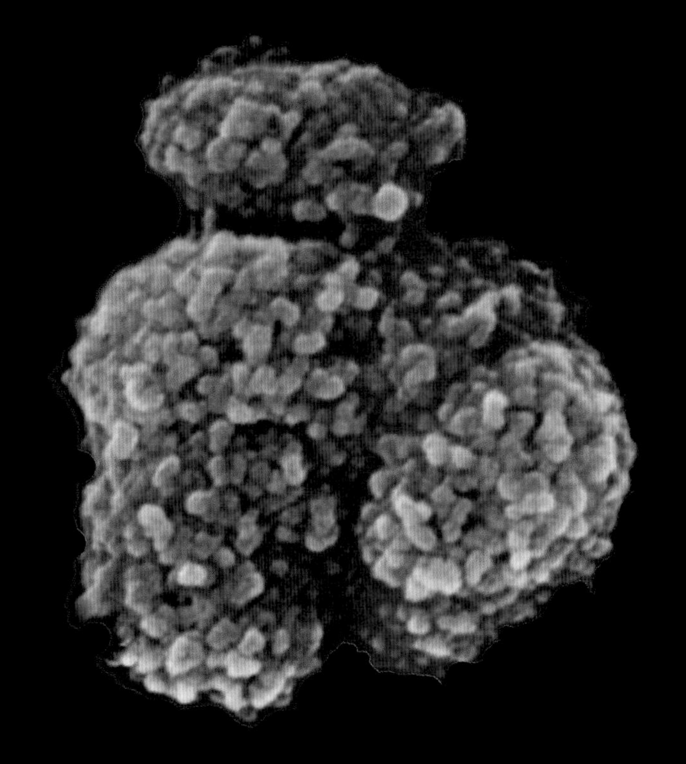

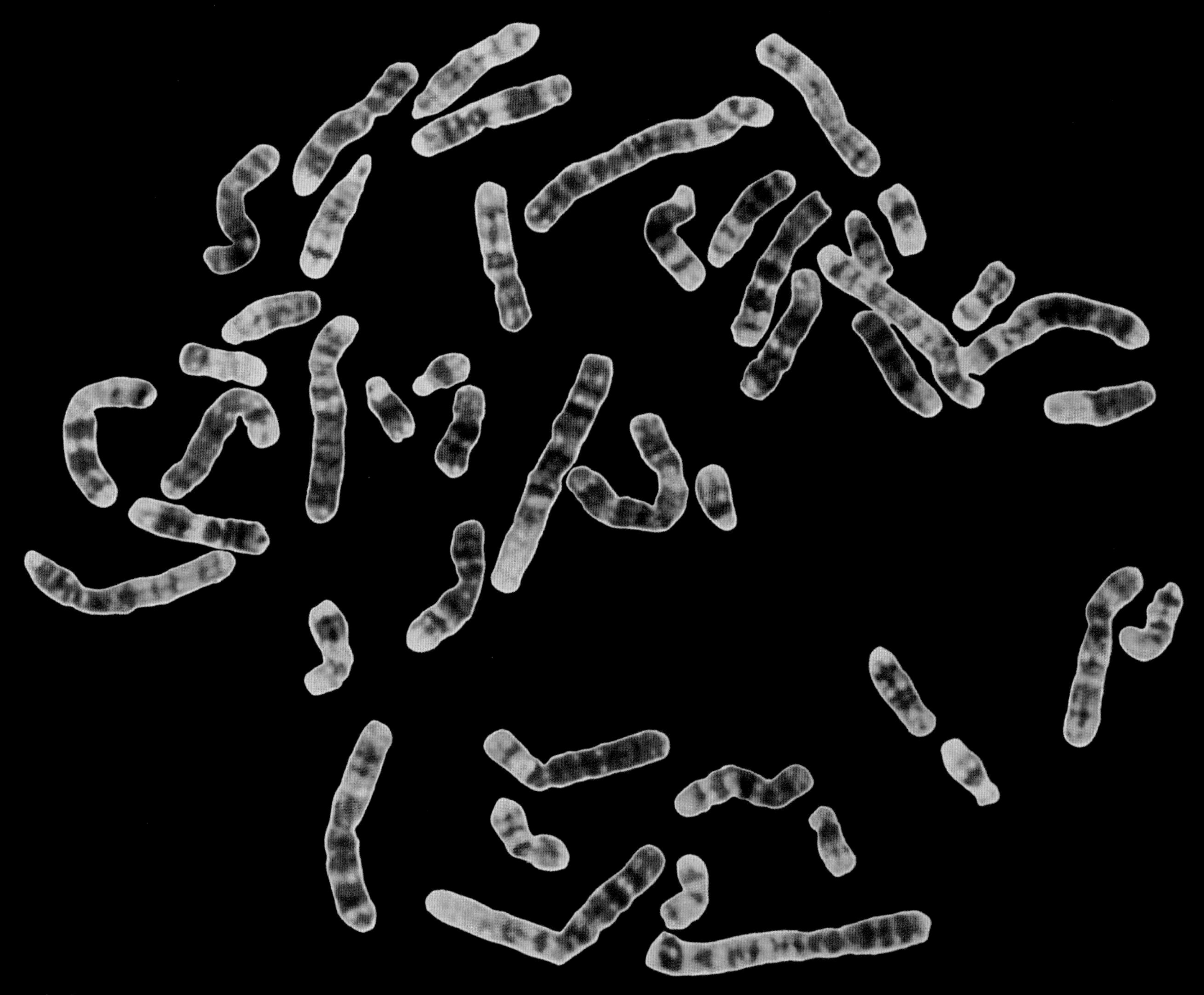

Male chromosomes (coloured light micrograph)
These 46 striped rods are the 46 chromosomes of a male human. Each chromosome has two arms; the bands on them are different genes, revealed by staining. Here the chromosomes are unpaired; pairing might be done by cutting up the image and pairing off each chromosome according to length, shape and banding pattern. Such a pairing is called a karyotype, and doctors use it to find particular genetic information and identify missing or abnormal chromosomes. (Magnification: x2400 at 6x7cm/2¼x2¾in size)

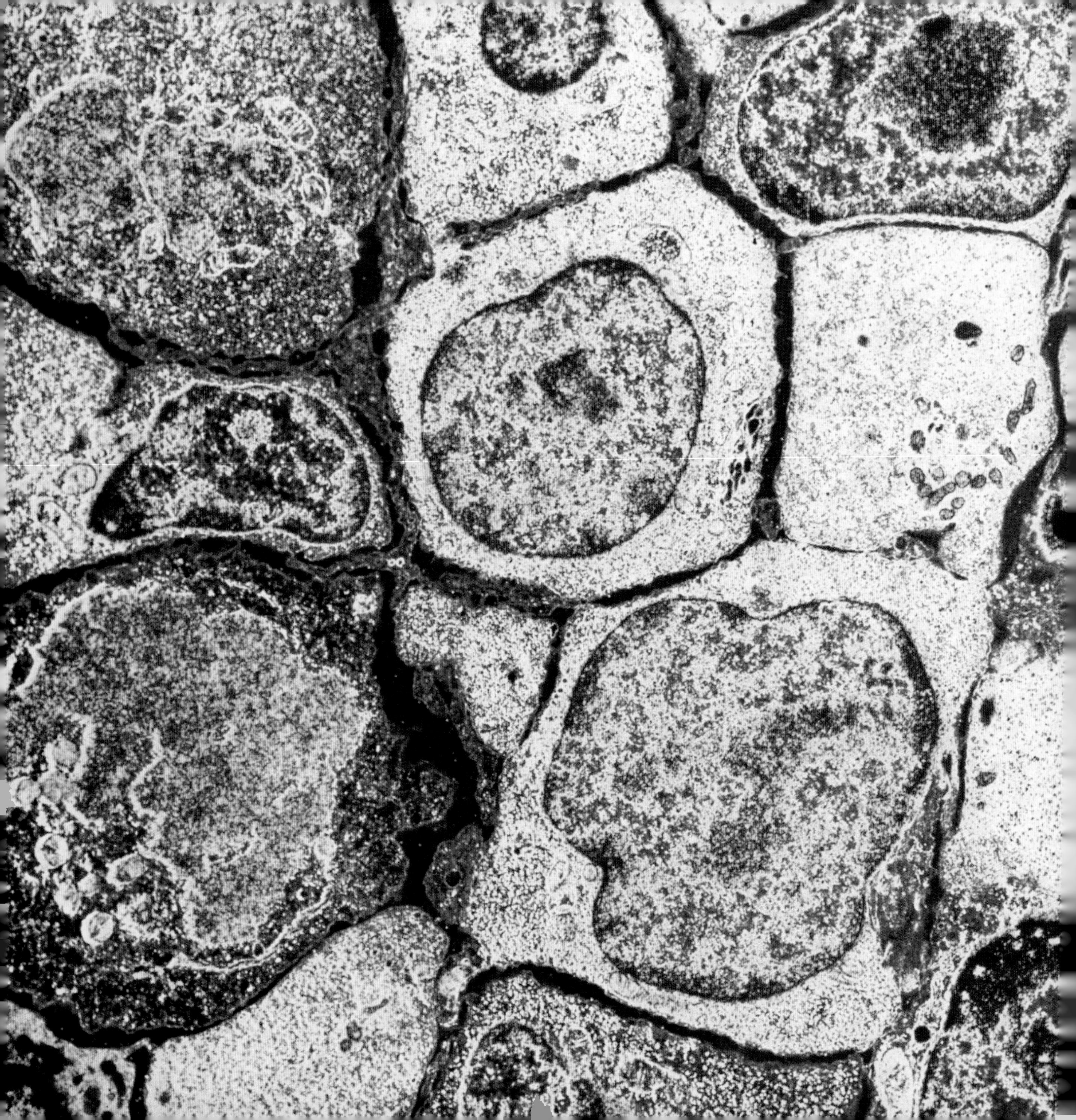

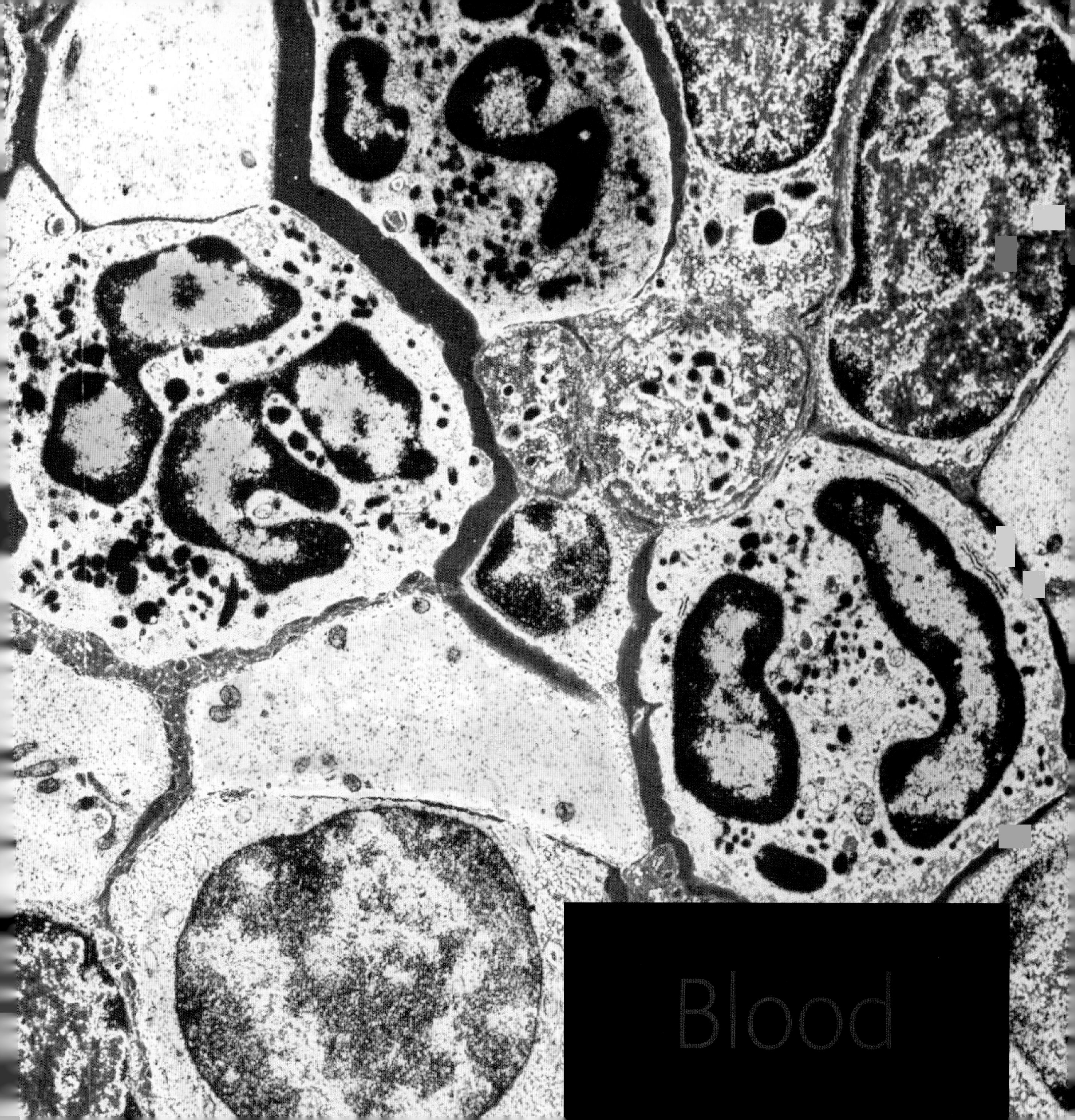

Blood

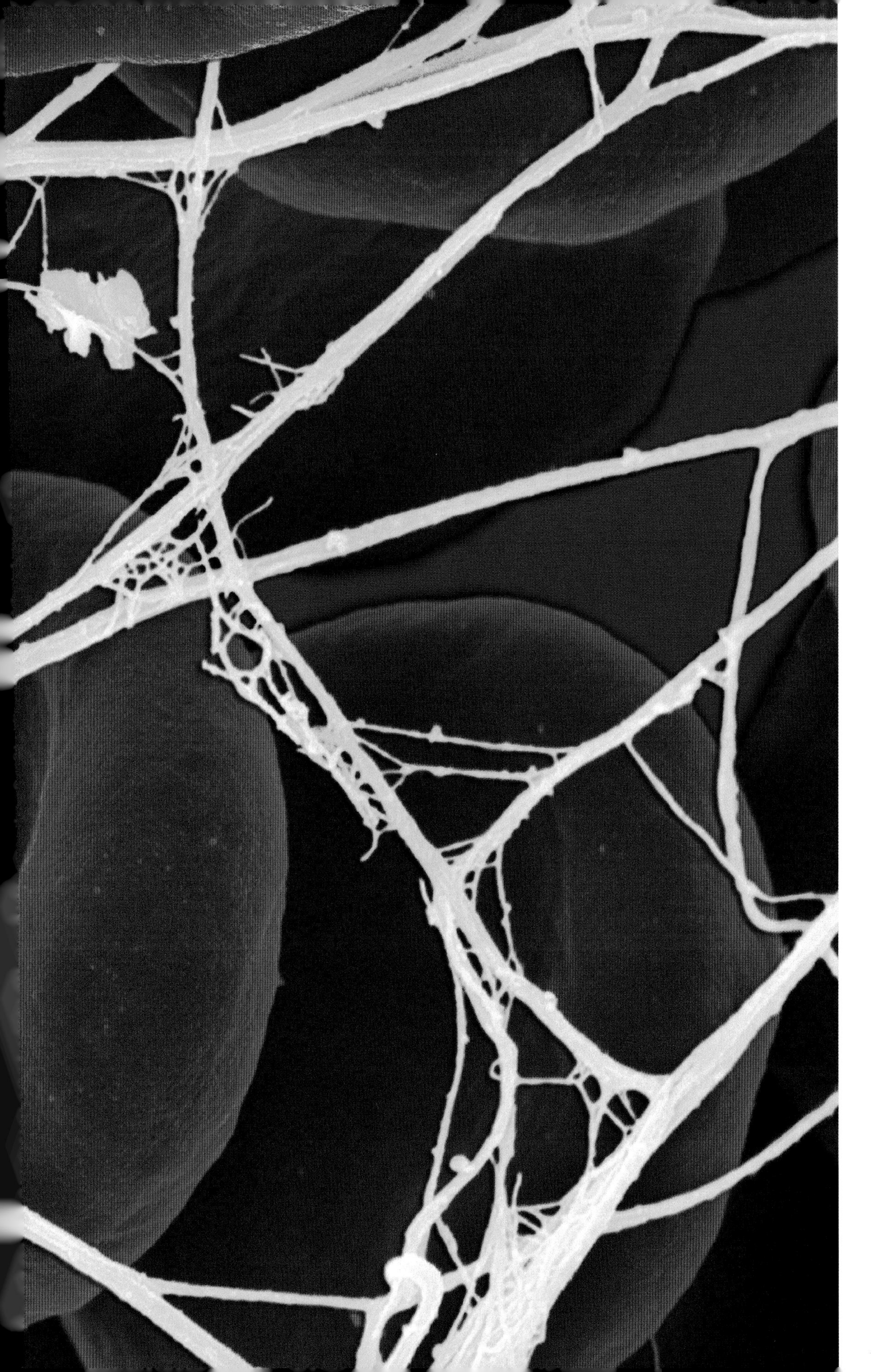

Previous page: Bone marrow (coloured transmission electron micrograph)
Blood cells have only a short lifespan – red ones about 120 days, some white ones as little as three. This necessitates their constant replacement, in a production process called haemopoiesis. They all derive from a single multi-purpose stem cell, generated in our bone marrow. Stem cells divide slowly before becoming one of five very different cell types: red cells (erythrocytes), white cells (lymphocytes, granulocytes or monocytes) and platelets (thrombocytes). This image shows those various transformations underway. (Magnification: x2300 at 5x7cm/2x2¾in size)

Left: Red blood cells in a blood clot (coloured scanning electron micrograph)
Alongside the red and white blood cells, blood also carries cells called platelets. Their function is to sense blood loss, for example as the result of a cut, and to plug the gap from within. Platelets release a protein called fibrin. The fibrin filaments criss-cross to form a crude but effective net, capturing the escaped red blood cells and forming a protective barrier across the gap on the outside. (Magnification: x7000 at 10x12cm/4x4¾in size)

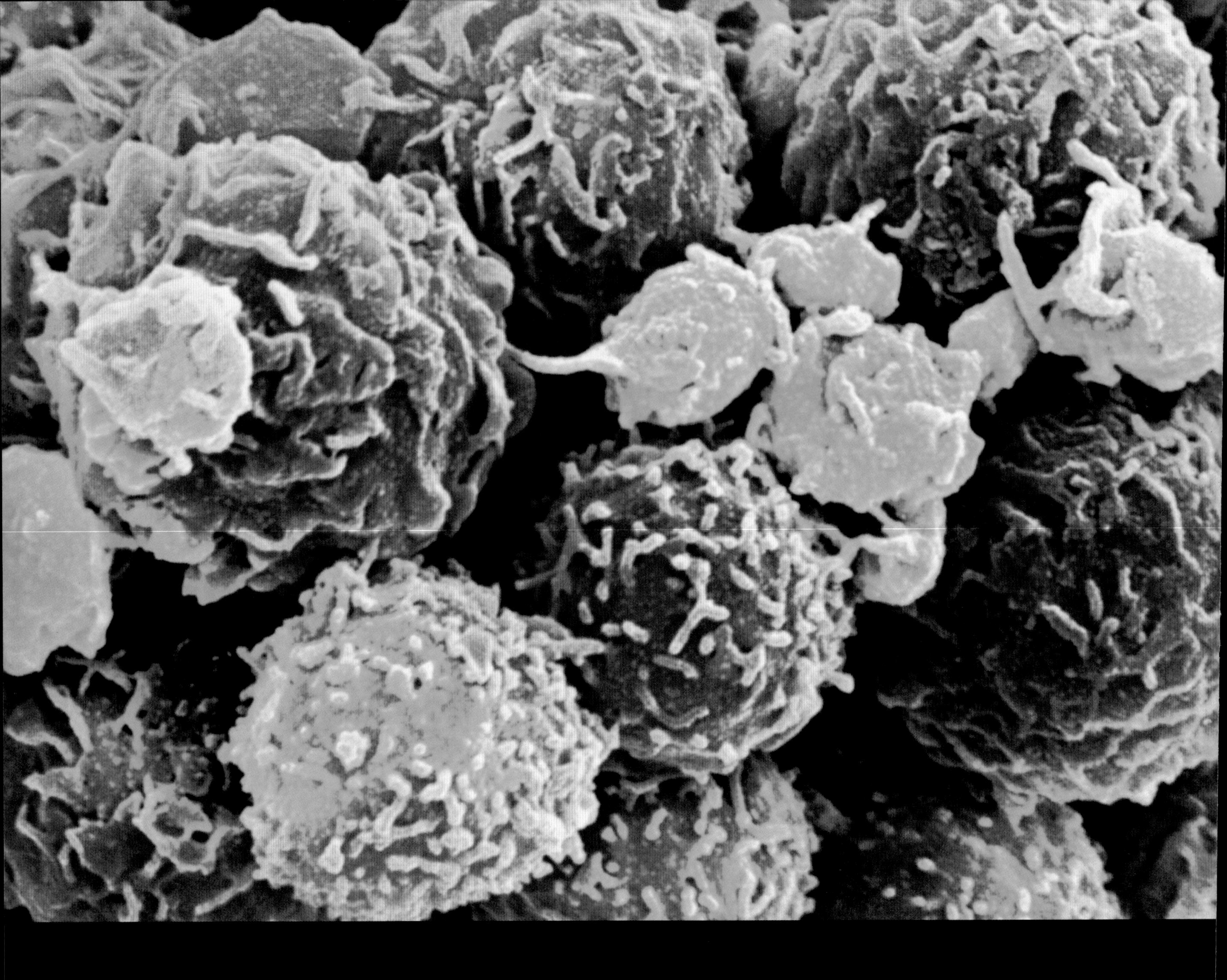

White blood cells and platelets (coloured scanning electron micrograph)
The larger, darker shapes in this blue cluster are white blood cells, the smaller, lighter ones platelets. Together these protective cells make up

Lymphocyte white blood cell (coloured scanning electron micrograph)
There are many kinds of white blood cells, and this lymphocyte is one of the commonest. The tiny tendrils improve its function by increasing its surface area. Two main types of lymphocytes occur in the human body. B cells mature in our bone marrow before entering the blood stream; they identify bacterial micro-organisms and then make antibodies. T cells migrate to the thymus gland and mature there; in the blood, they track down and help to actively destroy viral bodies. (Magnification: unknown)

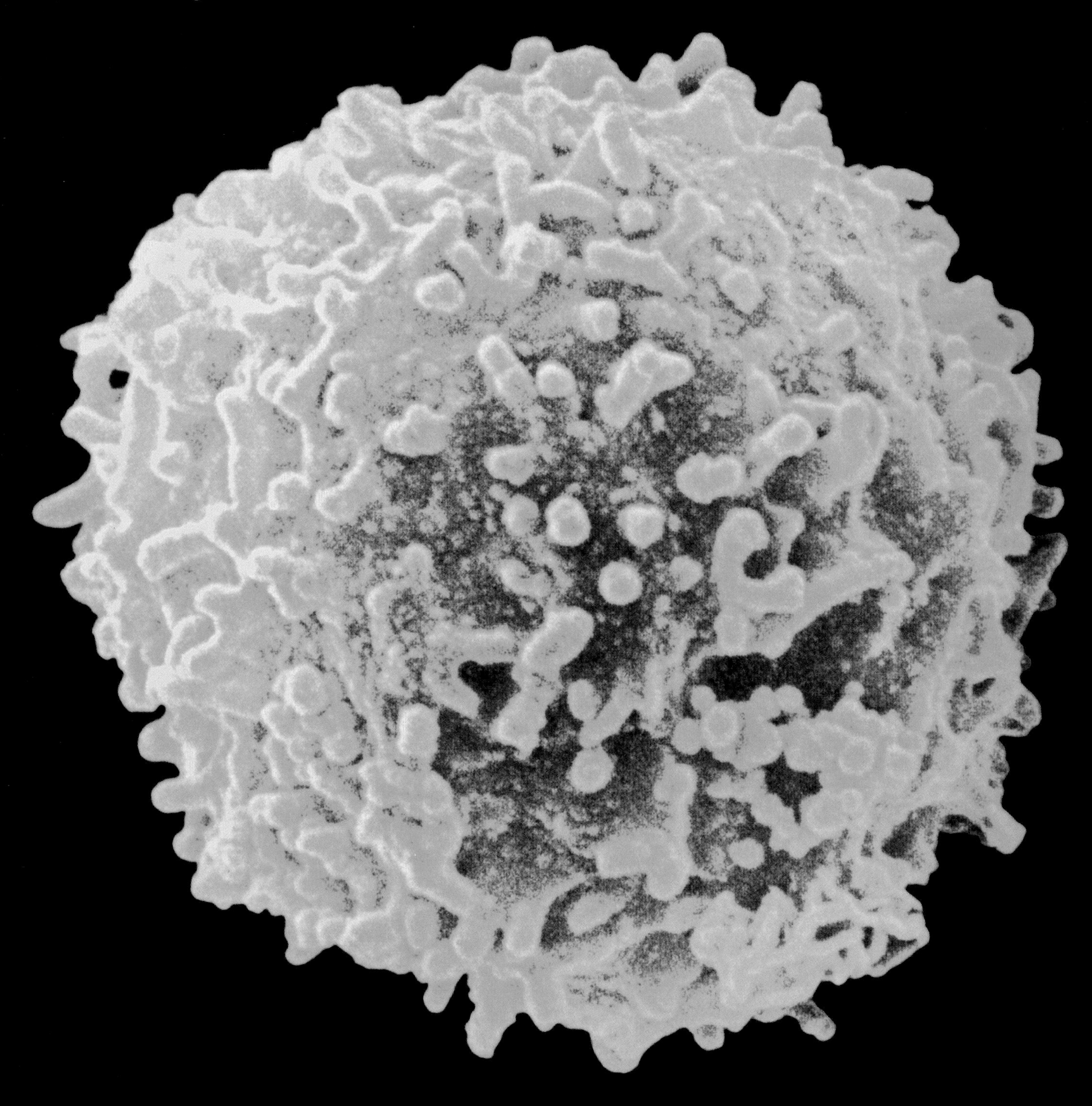

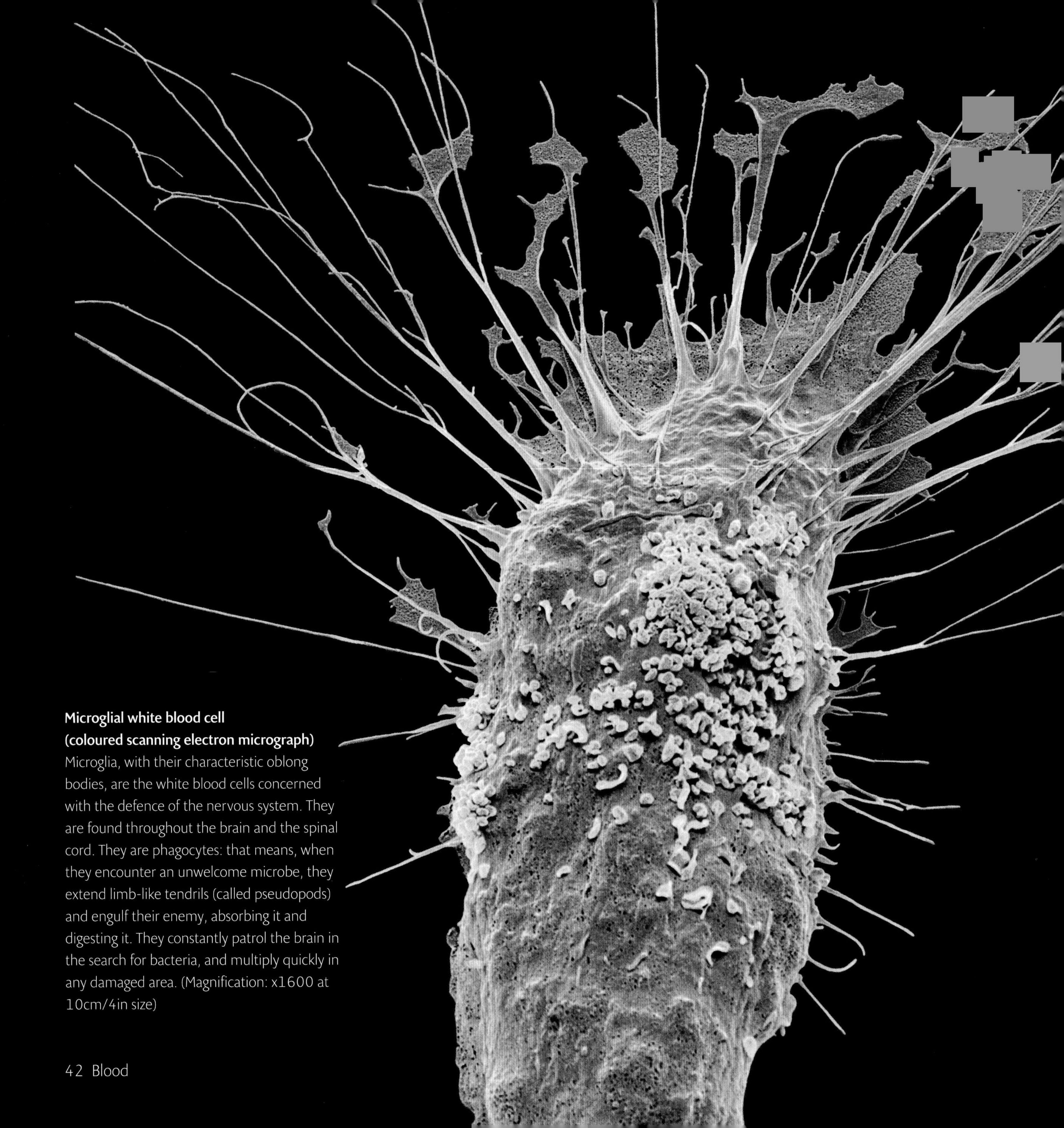

Microglial white blood cell (coloured scanning electron micrograph)
Microglia, with their characteristic oblong bodies, are the white blood cells concerned with the defence of the nervous system. They are found throughout the brain and the spinal cord. They are phagocytes: that means, when they encounter an unwelcome microbe, they extend limb-like tendrils (called pseudopods) and engulf their enemy, absorbing it and digesting it. They constantly patrol the brain in the search for bacteria, and multiply quickly in any damaged area. (Magnification: x1600 at 10cm/4in size)

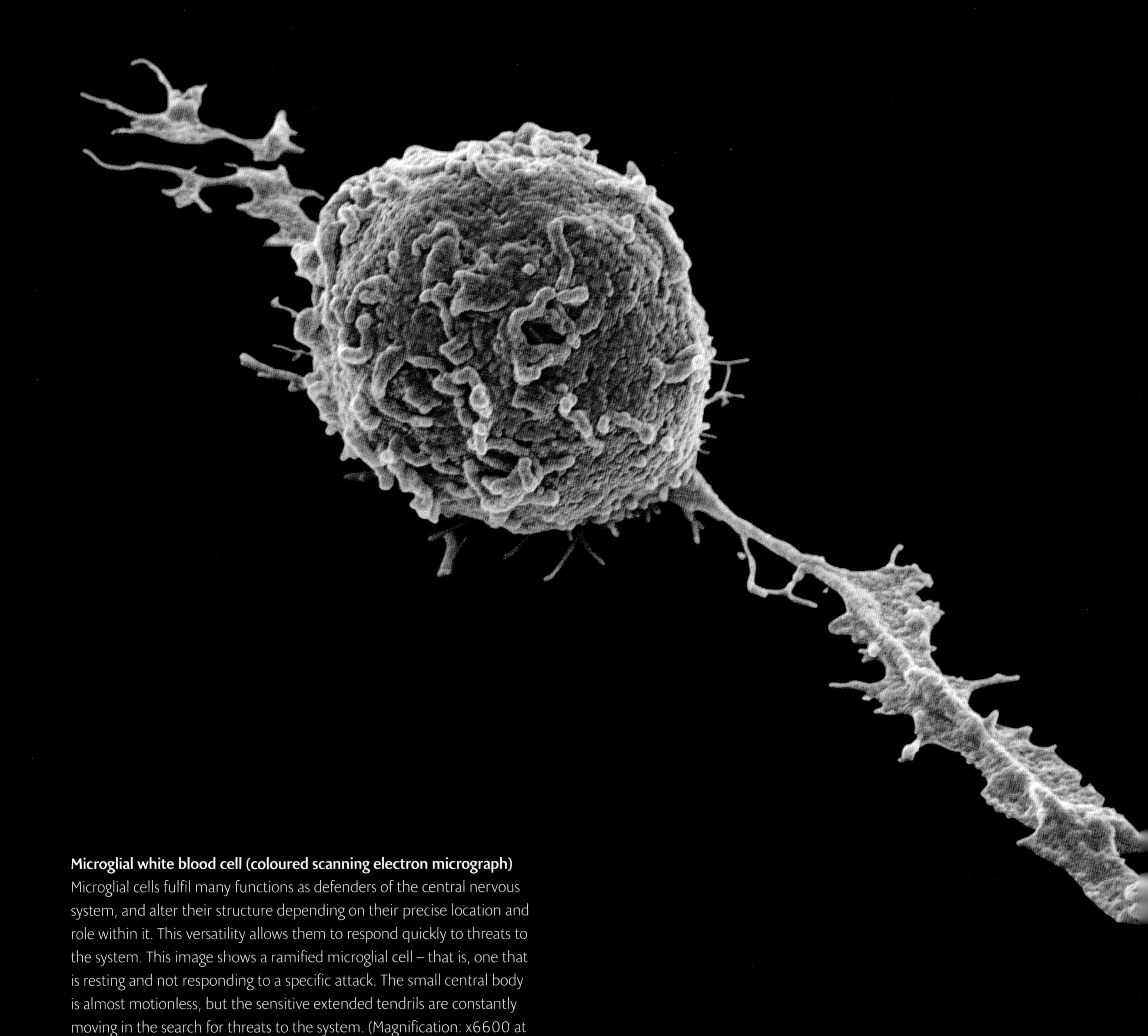

Microglial white blood cell (coloured scanning electron micrograph)
Microglial cells fulfil many functions as defenders of the central nervous system, and alter their structure depending on their precise location and role within it. This versatility allows them to respond quickly to threats to the system. This image shows a ramified microglial cell – that is, one that is resting and not responding to a specific attack. The small central body is almost motionless, but the sensitive extended tendrils are constantly moving in the search for threats to the system. (Magnification: x6600 at 10cm/4in size)

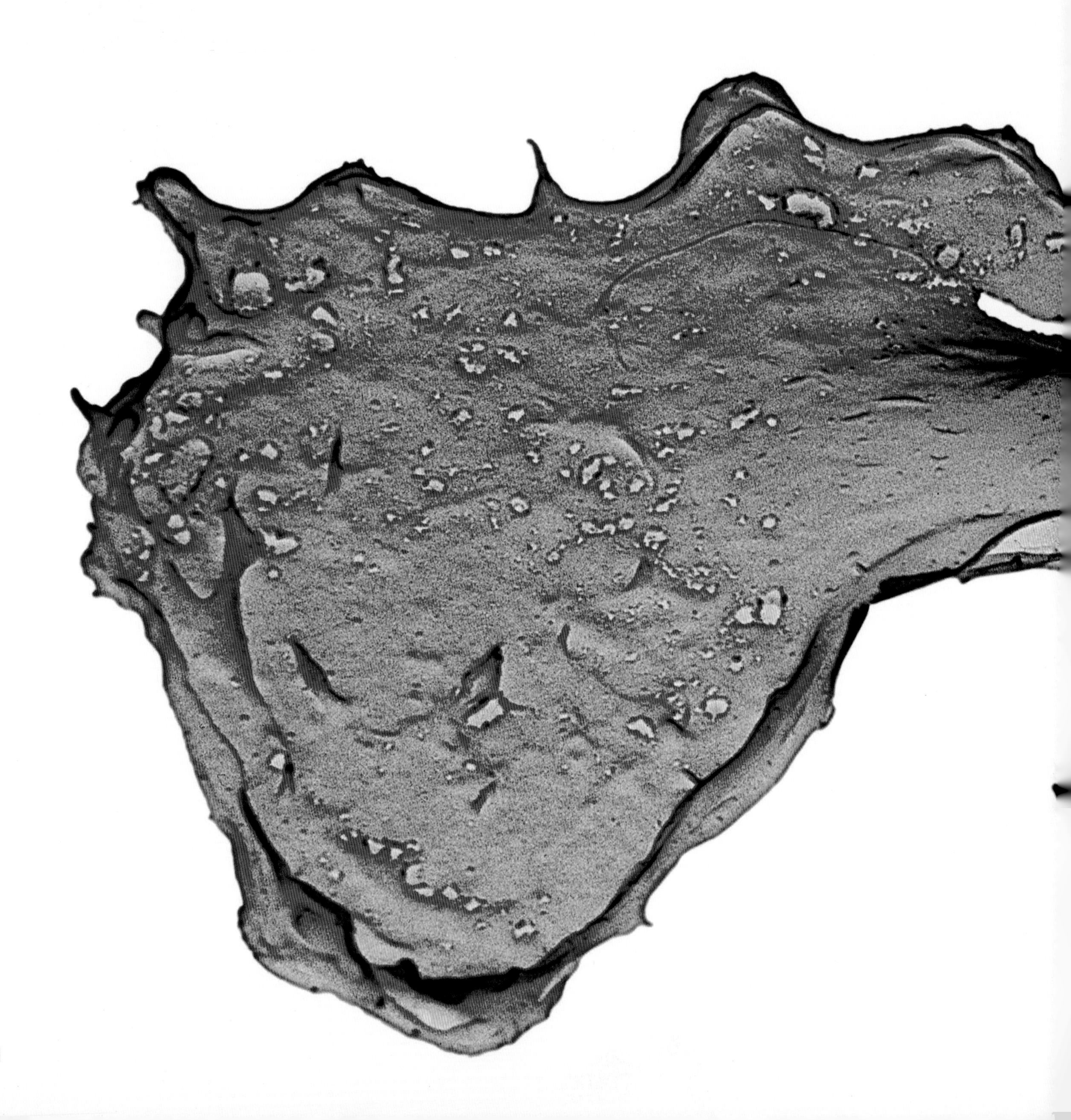

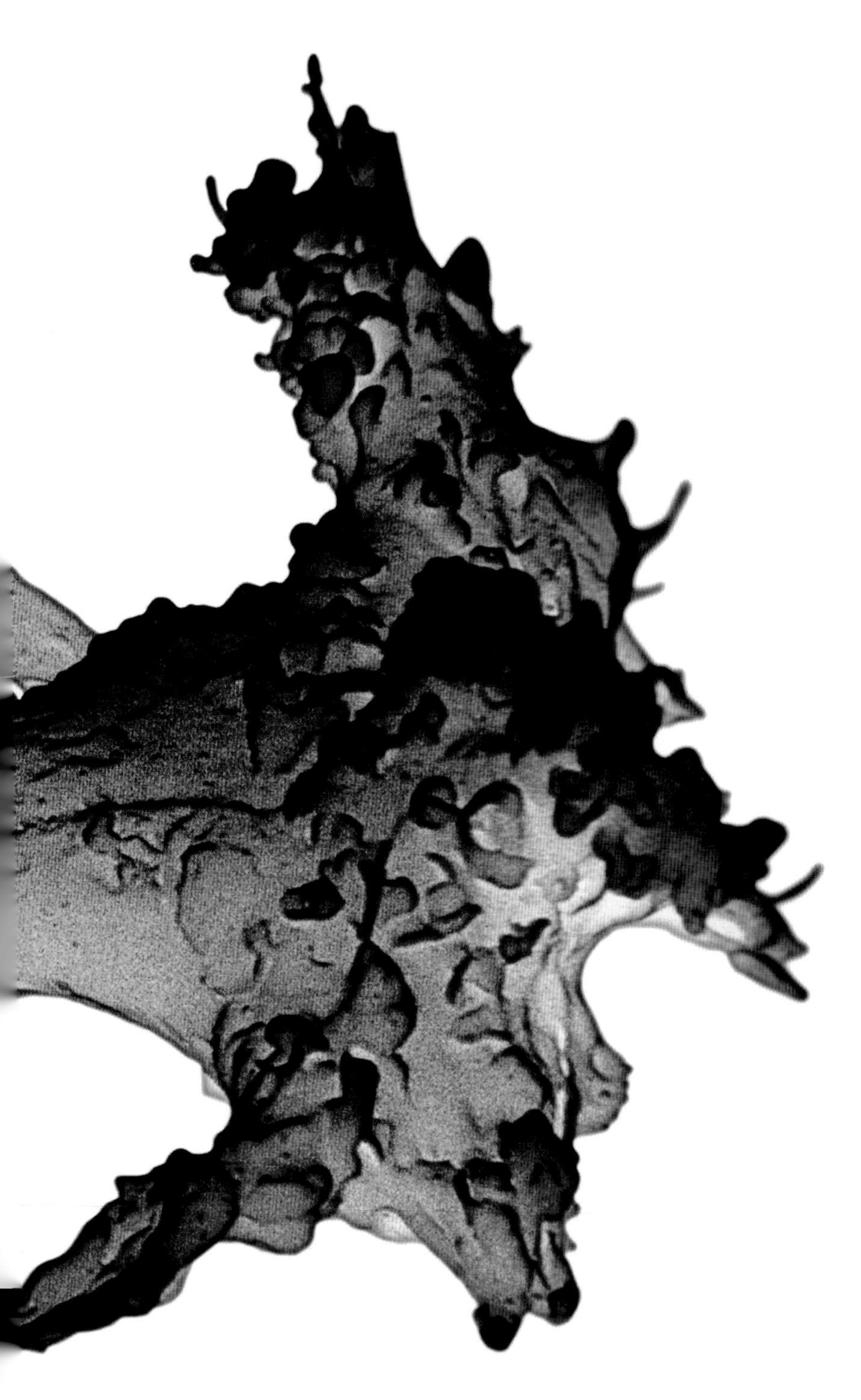

Macrophage with an extended pseudopod (coloured scanning electron micrograph)

A macrophage is normally a rounded white blood cell (the red portion in this image). Its job is to find a source of contamination – for example a pathogen, a dead cell or some other cellular debris – in the blood supply. When it does, it develops crude limbs, called pseudopods (the purple portion), which it extends to engulf the offending object. The macrophage then draws the object inside itself and there neutralizes or eliminates it. (Magnification: unknown)

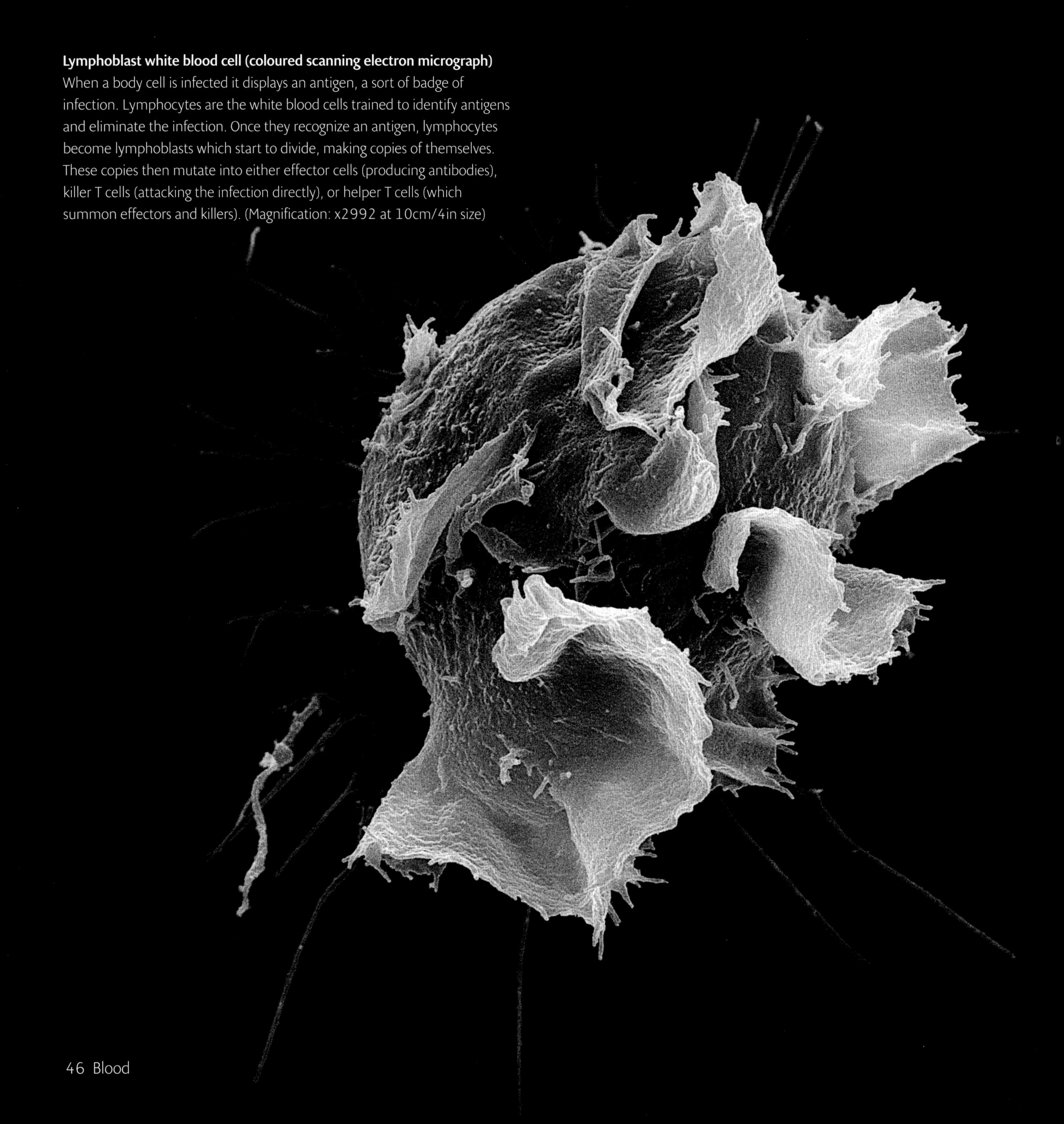

Lymphoblast white blood cell (coloured scanning electron micrograph)
When a body cell is infected it displays an antigen, a sort of badge of infection. Lymphocytes are the white blood cells trained to identify antigens and eliminate the infection. Once they recognize an antigen, lymphocytes become lymphoblasts which start to divide, making copies of themselves. These copies then mutate into either effector cells (producing antibodies), killer T cells (attacking the infection directly), or helper T cells (which summon effectors and killers). (Magnification: x2992 at 10cm/4in size)

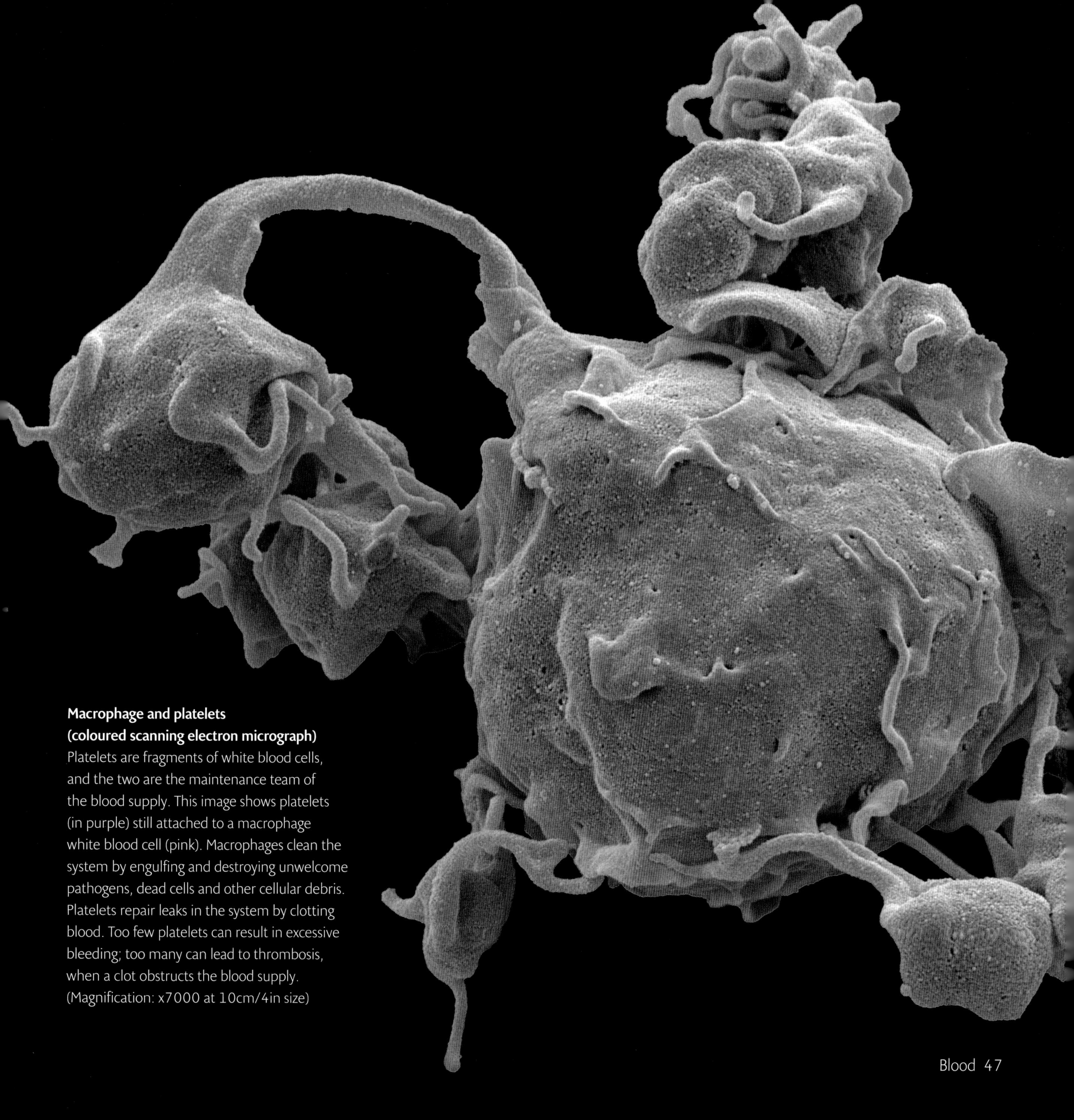

Macrophage and platelets (coloured scanning electron micrograph)
Platelets are fragments of white blood cells, and the two are the maintenance team of the blood supply. This image shows platelets (in purple) still attached to a macrophage white blood cell (pink). Macrophages clean the system by engulfing and destroying unwelcome pathogens, dead cells and other cellular debris. Platelets repair leaks in the system by clotting blood. Too few platelets can result in excessive bleeding; too many can lead to thrombosis, when a clot obstructs the blood supply. (Magnification: x7000 at 10cm/4in size)

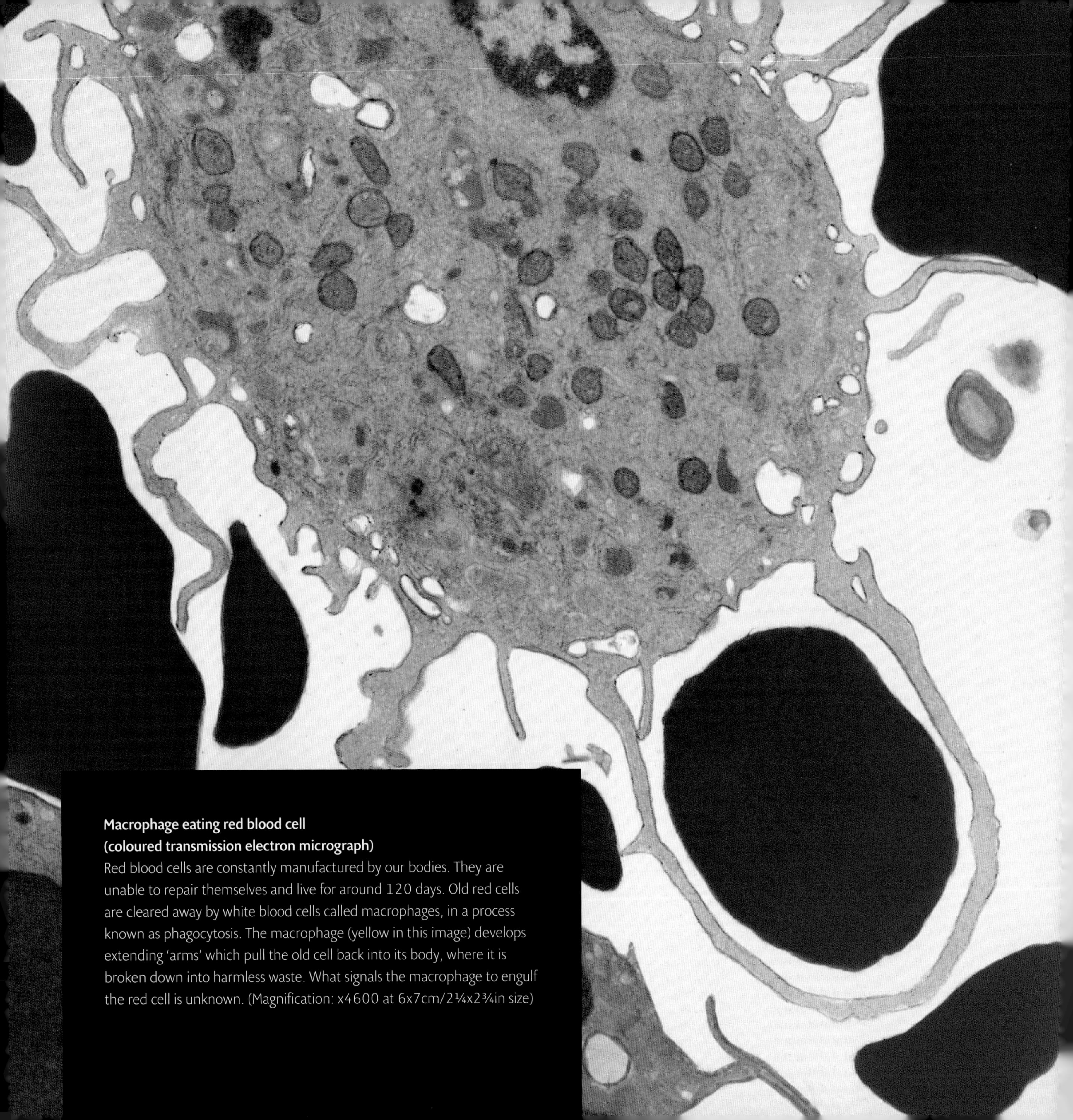

Macrophage eating red blood cell (coloured transmission electron micrograph)

Red blood cells are constantly manufactured by our bodies. They are unable to repair themselves and live for around 120 days. Old red cells are cleared away by white blood cells called macrophages, in a process known as phagocytosis. The macrophage (yellow in this image) develops extending 'arms' which pull the old cell back into its body, where it is broken down into harmless waste. What signals the macrophage to engulf the red cell is unknown. (Magnification: x4600 at 6x7cm/2¼x2¾in size)

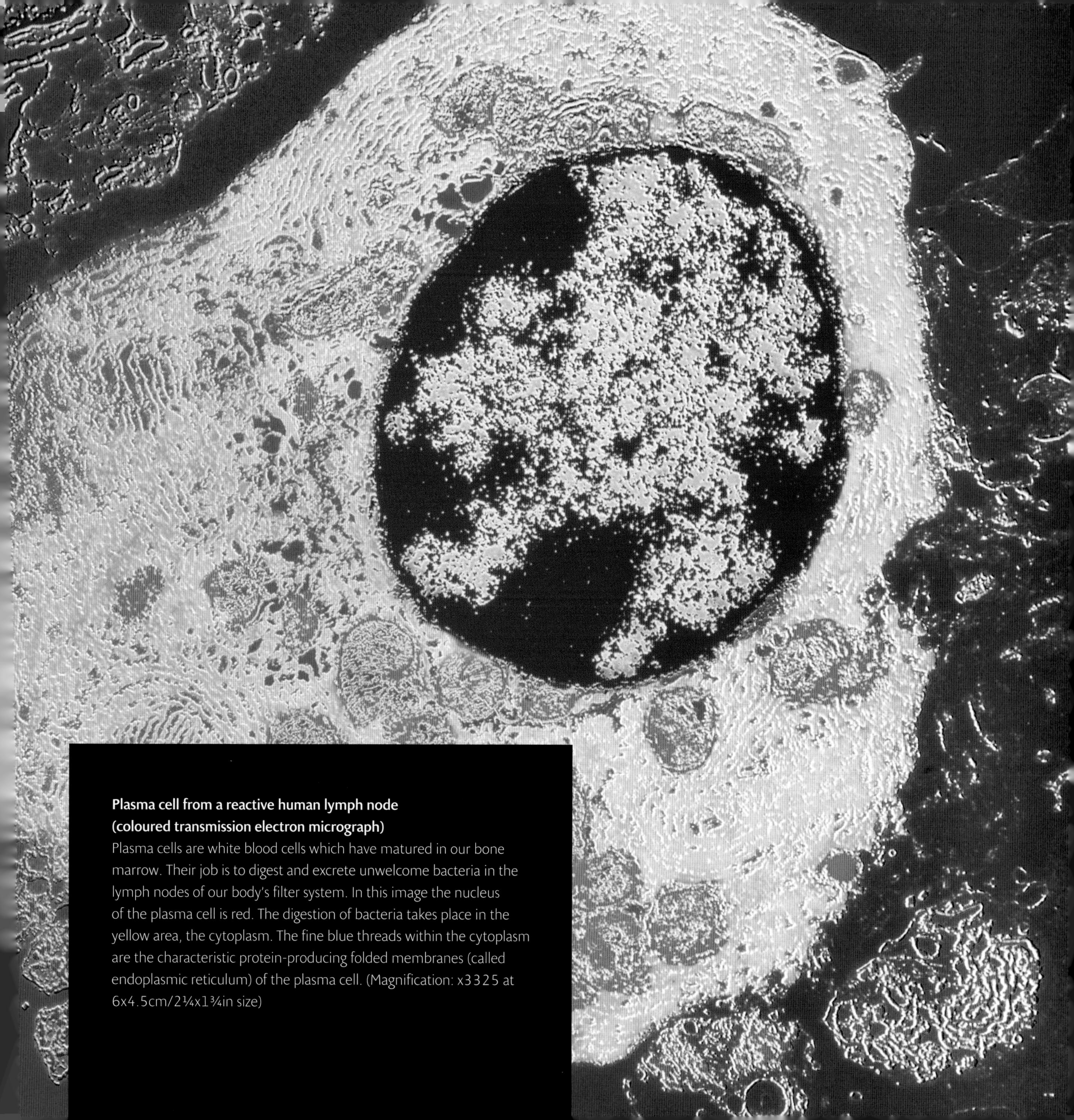

Plasma cell from a reactive human lymph node (coloured transmission electron micrograph)
Plasma cells are white blood cells which have matured in our bone marrow. Their job is to digest and excrete unwelcome bacteria in the lymph nodes of our body's filter system. In this image the nucleus of the plasma cell is red. The digestion of bacteria takes place in the yellow area, the cytoplasm. The fine blue threads within the cytoplasm are the characteristic protein-producing folded membranes (called endoplasmic reticulum) of the plasma cell. (Magnification: x3325 at 6x4.5cm/2¼x1¾in size)

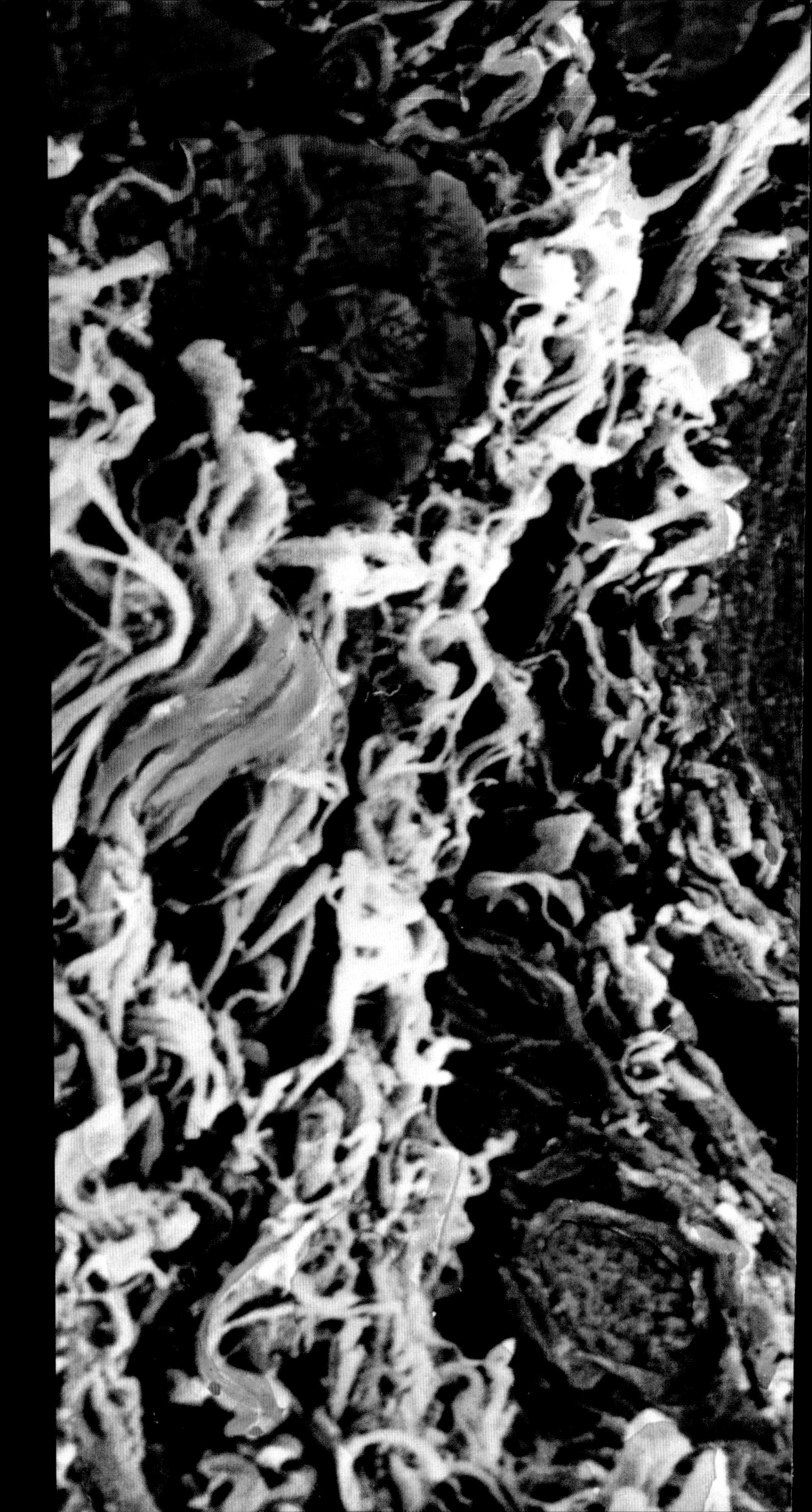

Cross-section of an artery (coloured scanning electron micrograph)

Our arteries consist of many layers. The inner wall, closest to the dark passage through which blood flows, is the tunica intima (purple), here seen with folds because the artery has contracted. Outside that is a layer of smooth muscle (dark brown), the tunica media; and beyond that a sheath of flexible material (light brown), the tunica adventitia. This particular artery supplies the muscle tissue (red) of the tongue. The small blue oval below the artery is a nerve. (Magnification: x170 at 6x7cm/2¼x2¾in size)

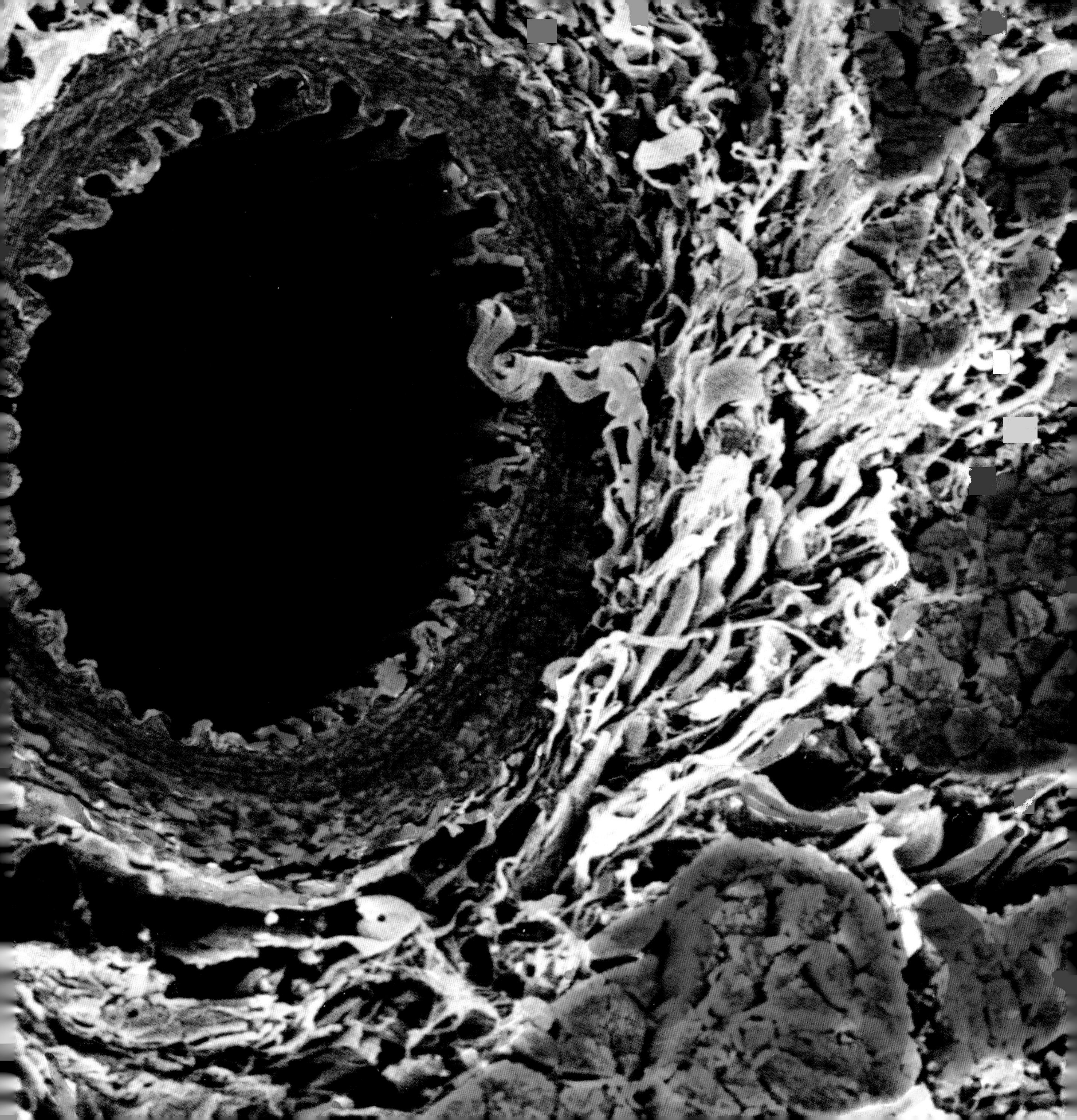

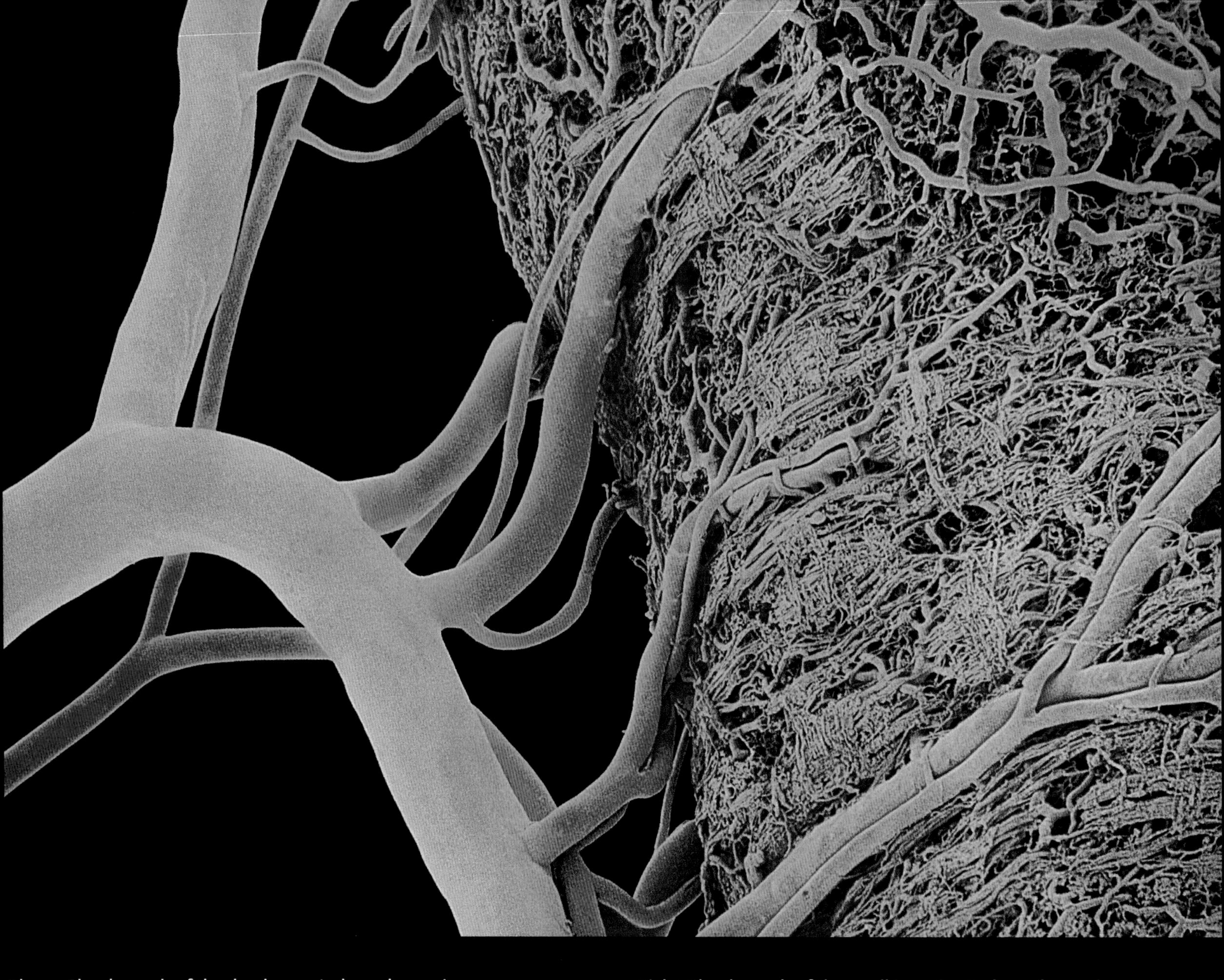

Above: Blood vessels of the duodenum (coloured scanning electron micrograph)

This branching network of fine vessels threads through the tissue of the duodenum, the first section of the small intestine into which food passes from the stomach. The finest vessels of all, the capillaries, have permeable walls which allow gases and nutrients to pass through them from the blood to the surrounding tissue. This image was made by injecting resin into the blood vessels. Once the resin has hardened, the surrounding tissue is chemically removed to reveal the resin cast. (Magnification: x32 at 6x7cm/2¼x2¾in size)

Right: Blood vessels of the small intestine (coloured scanning electron micrograph)

A resin cast of the blood supply to the tightly packed folds of the small intestine. The small intestine is the part of the digestive system in which nutrients are extracted from food. It consists of three sections: the duodenum, the jejunum and the ileum, from which waste passes to the large intestine. The average length of the human small intestine is about 7m (23ft) – slightly longer in women than in men. (Magnification: x30 at 6x7cm/2¼x2¾in size)

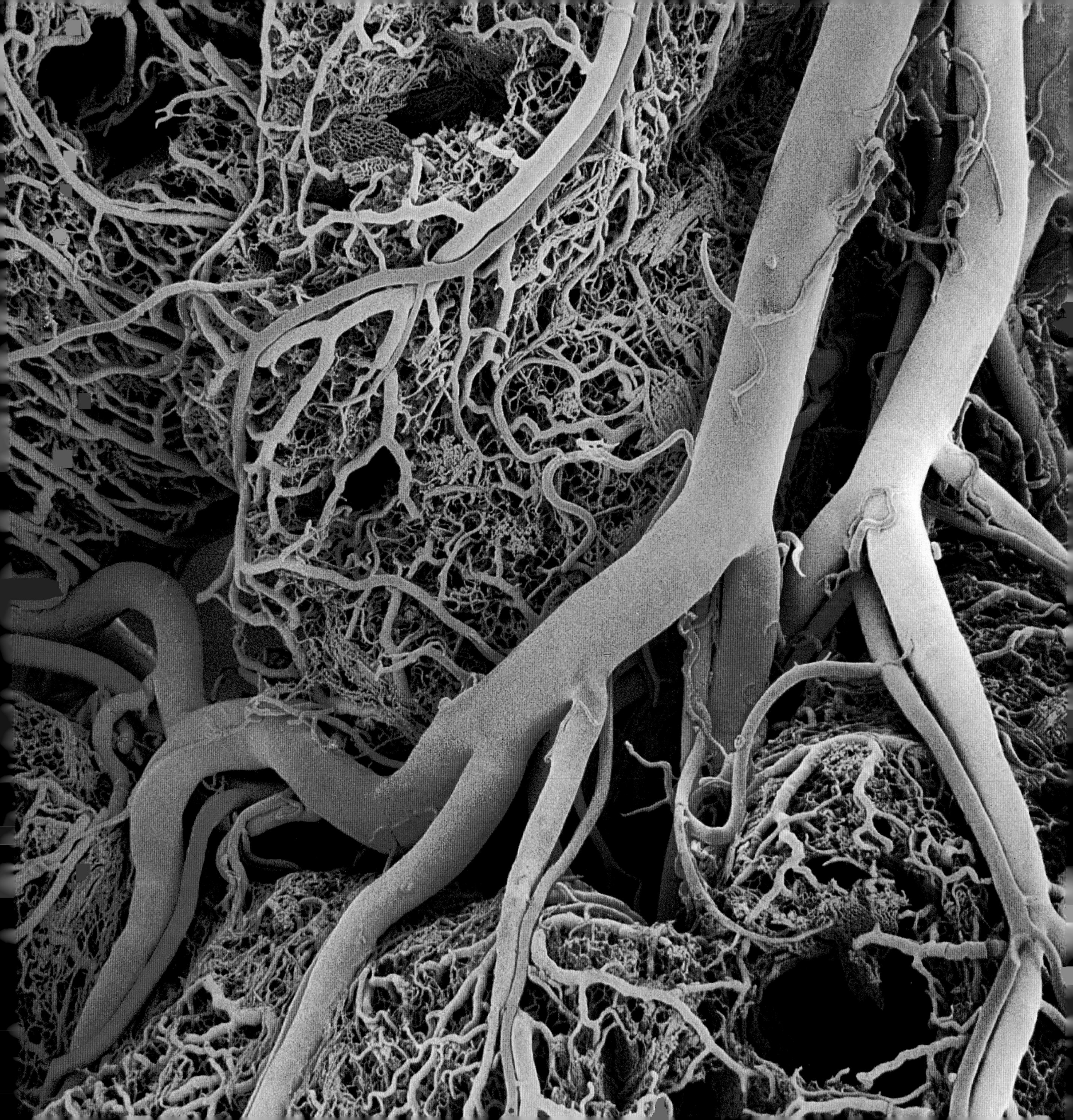

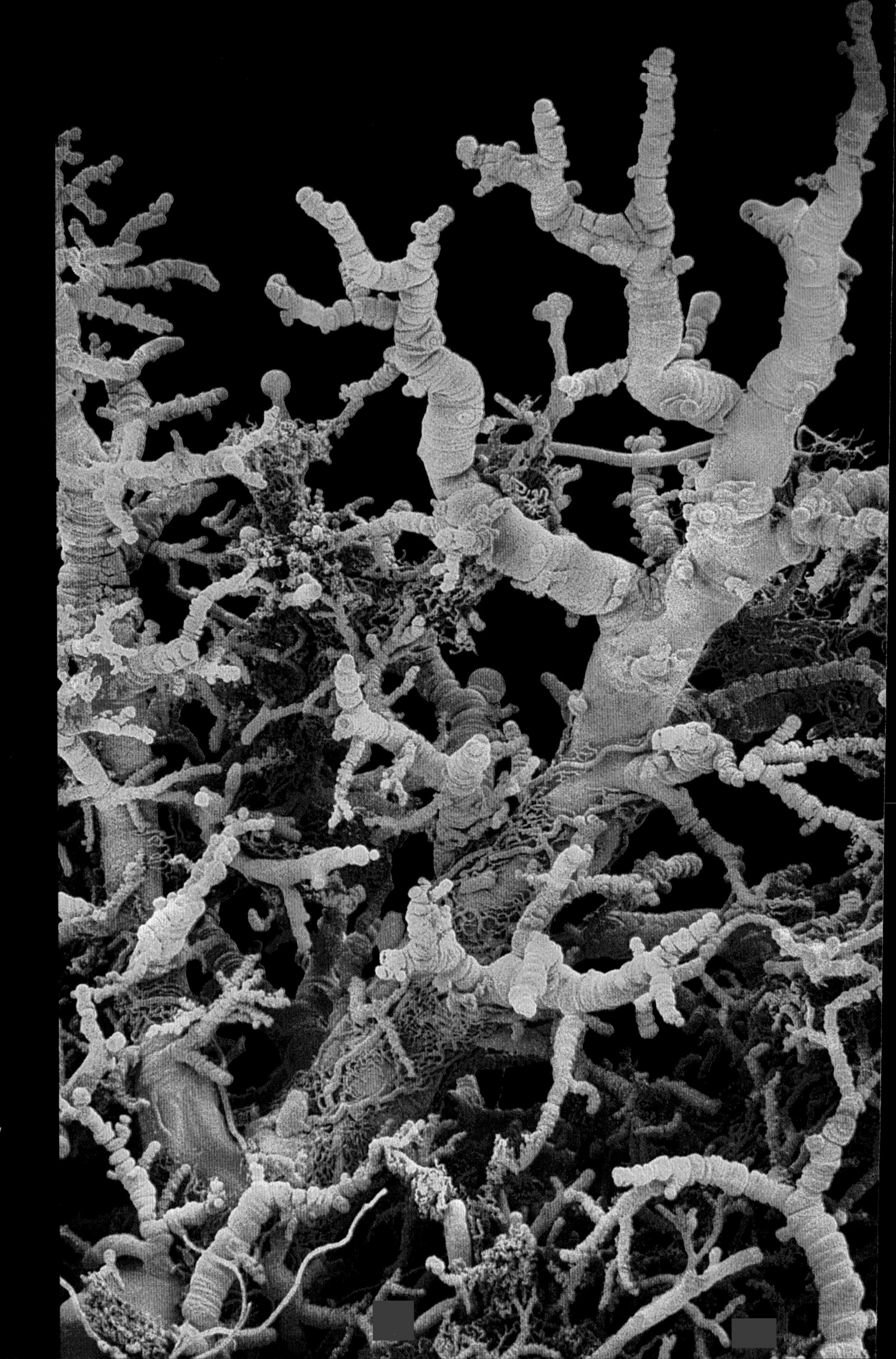

Blood vessels in a lung (coloured scanning electron micrograph)

Lungs breathe in vital oxygen to be delivered to the rest of the body, and breathe out carbon dioxide formed by cell respiration. It is the circulation of blood, seen here in a resin cast, which delivers both gases. In the lungs, pulmonary arteries deliver deoxygenated blood to air sacs called alveoli. Alveoli are covered in very fine capillaries – blood vessels with permeable walls which allow the exchange of gases. Pulmonary veins then return the revitalized blood to the system. (Magnification: x10 at 6x7cm/2¼x2¾in size)

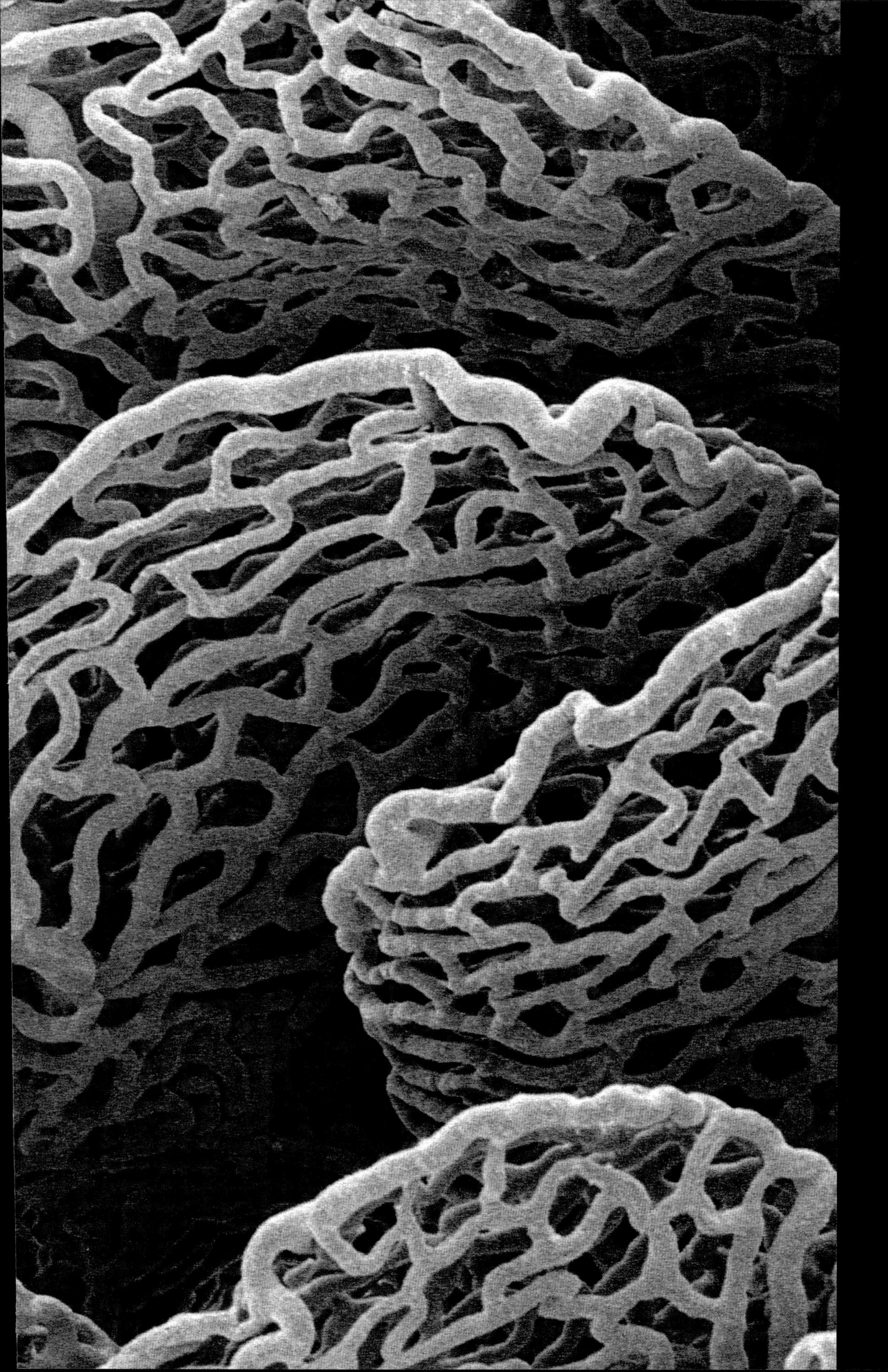

Blood vessels in the villi of the duodenum (coloured scanning electron micrograph)
Villi are finger-like projections on the inner wall of the intestines, which serve to increase the surface area available for the absorption of nutrients. Each 'finger' is richly supplied with blood vessels into which the nutrients are absorbed. In this image the vessels have been injected with resin; and once the resin has set, the surrounding tissues have been dissolved chemically, leaving a resin cast which allows us to see the trellis-like network of vessels clearly. (Magnification: x285 at 10cm/4in size)

Right: Brain blood vessels (coloured scanning electron micrograph)
A good supply of oxygen to the brain is essential – some brain cells die after only five minutes without oxygen. It is delivered by the vertebral arteries and the internal carotid artery, which divide into branches supplying the front, middle and rear brains and the cerebellum. Veins carry deoxygenated blood back towards the heart, coming together as they leave the brain in the internal jugular vein. In this image, veins are pink and arteries are orange. (Magnification: unknown)

Far right: Stomach wall blood vessels (coloured scanning electron micrograph)
The stomach breaks food down with a hard-hitting cocktail of enzymes to digest protein and hydrochloric acid to destroy bacteria. To protect itself from these powerful processes its wall is covered in a barrier of mucus produced by cardiac glands. In this resin cast the glands themselves have been removed to reveal a yellow web, the network of blood vessels which surround and supply them. The folds of the stomach allow it to expand as it fills. (Magnification: unknown)

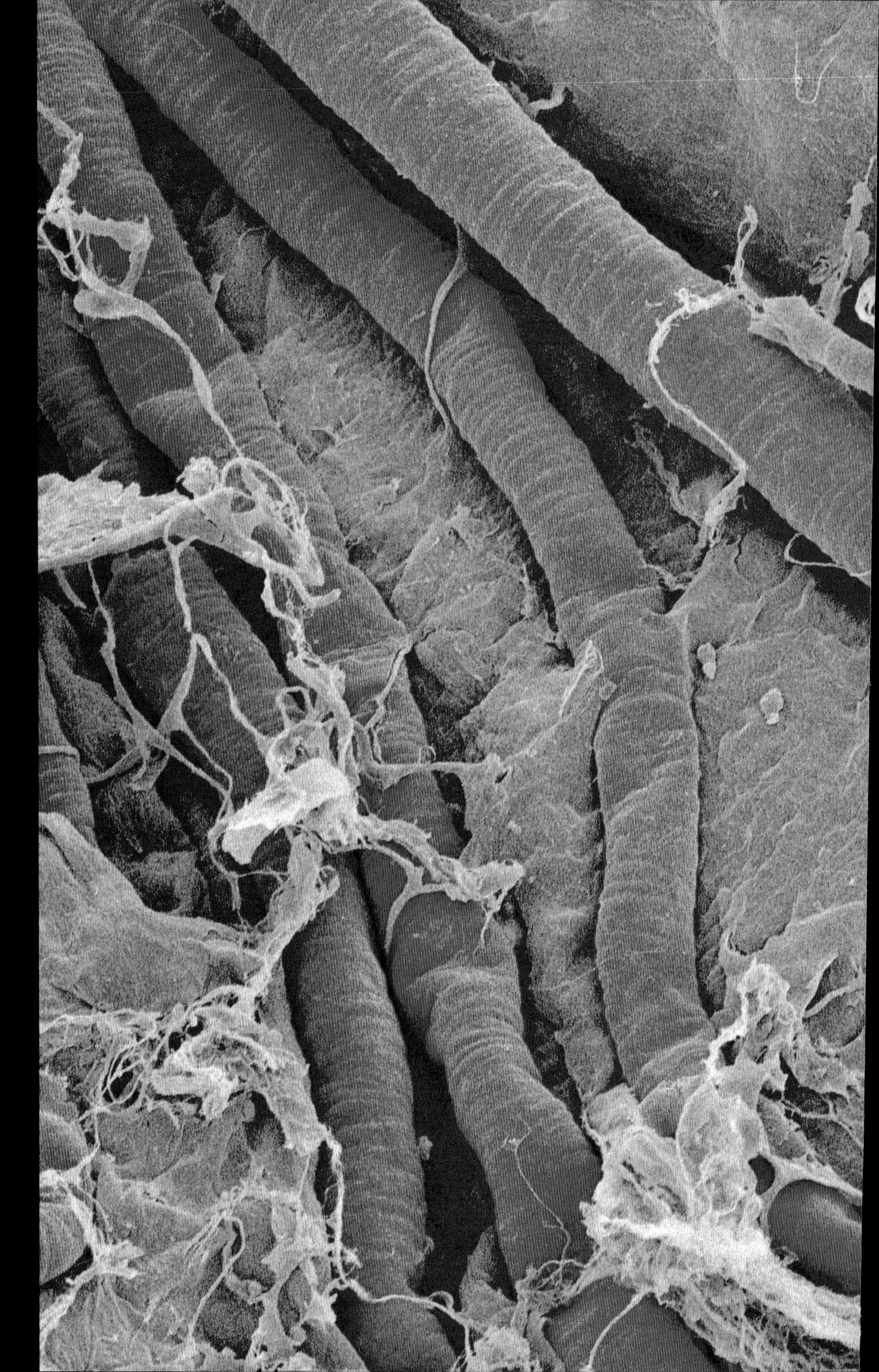

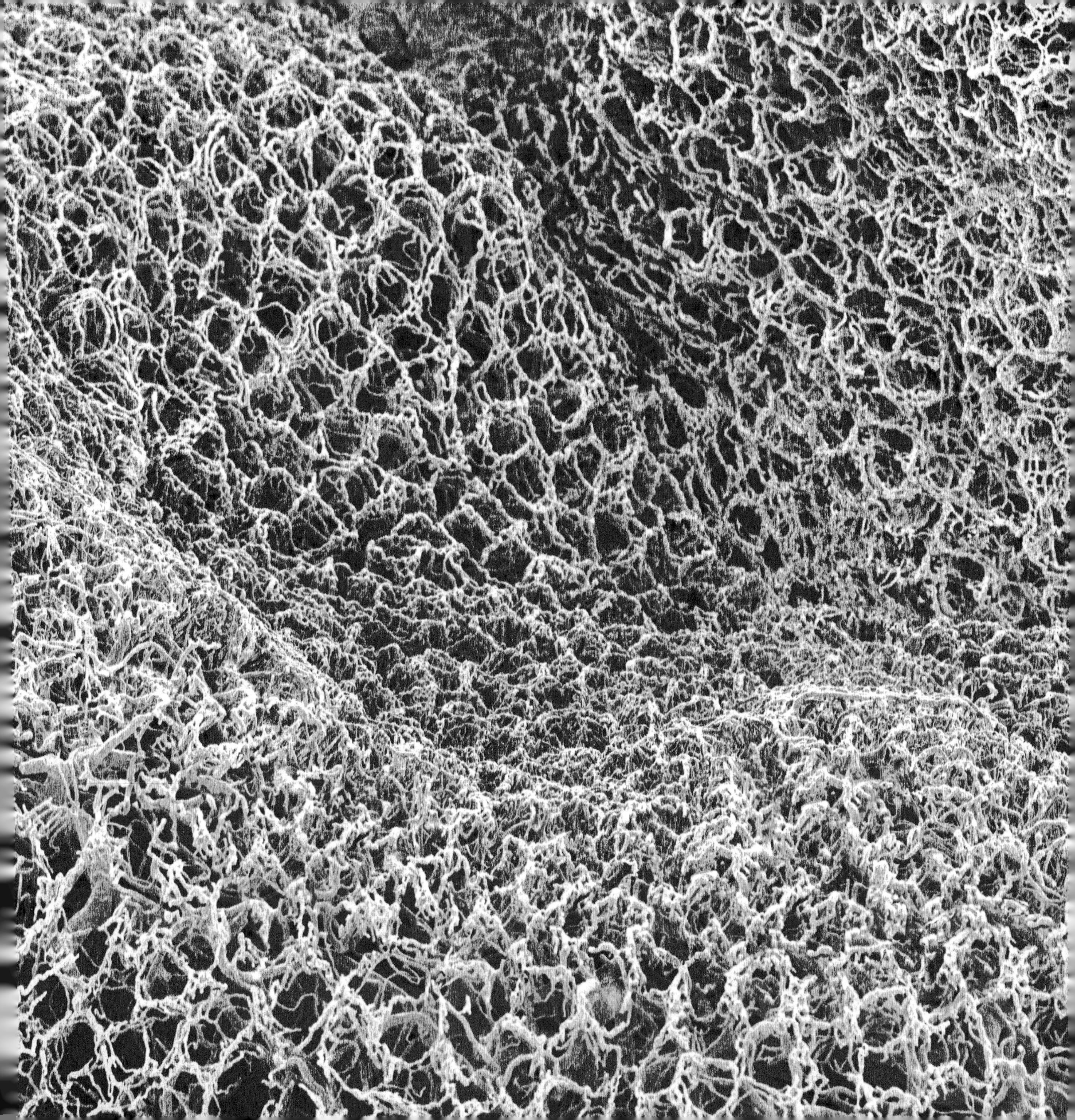

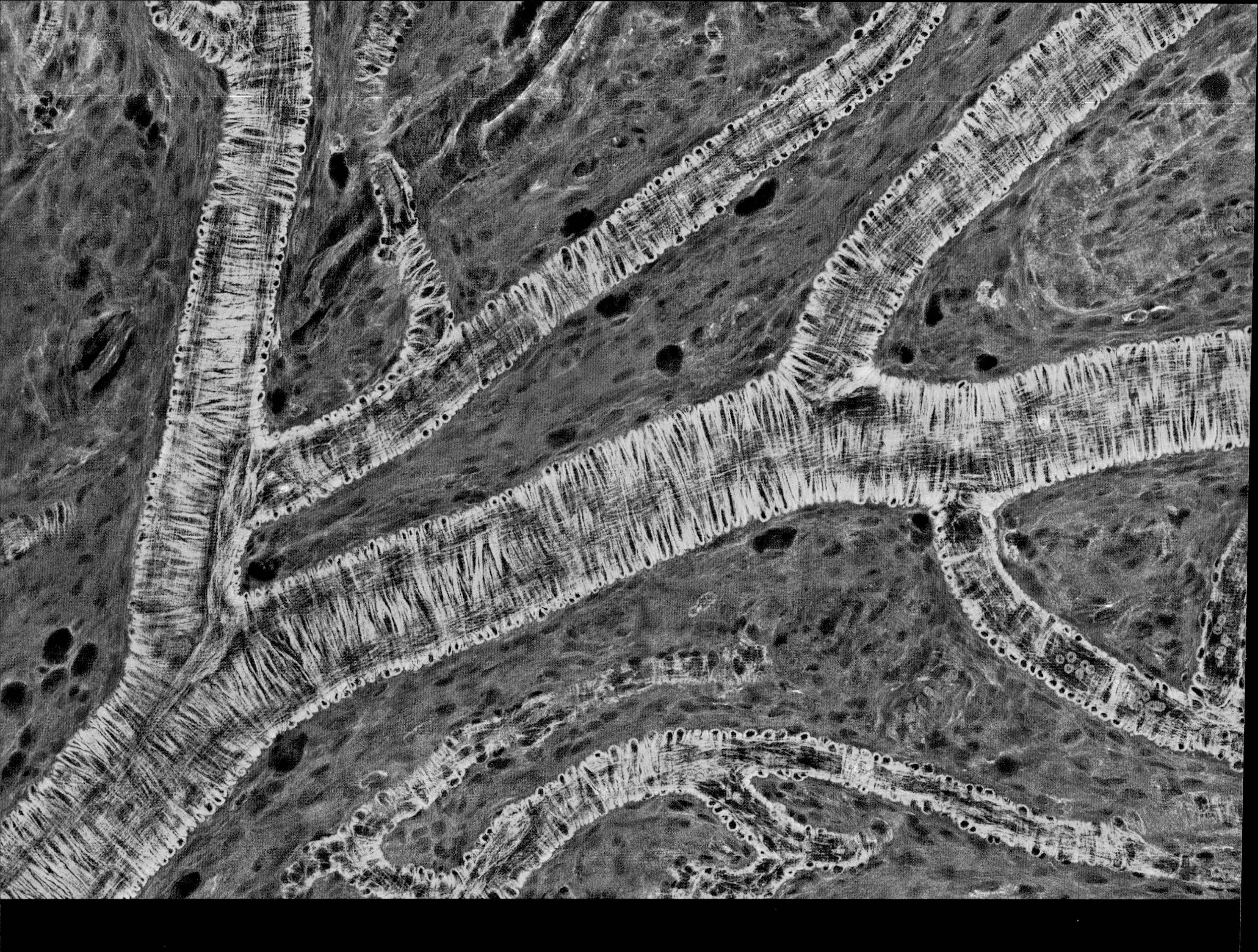

Above: Retina blood vessels (fluorescence light micrograph)
The retina is the light-sensitive membrane that lines the back of the eyeball. The blood vessels which supply the retina are encased in muscle which contracts and relaxes to control the flow. In this image a green fluorescent dye has been used to highlight strands of the protein actin, which plays a part in muscle contraction. Within some branches of the blood supply you can clearly see red blood cells, here passing from bottom to top and left to right. (Magnification: x200 at 10cm/4in size)

Right: Blood-brain barrier (confocal light micrograph)
In this image of a blood vessel (the dark channel) running through the brain, its inner walls are more densely packed with neurons (nerve cells, red) than elsewhere in the body. Surrounding the blood vessel are glial cells (green), which provide structural support and nutrition for neurons, and help maintain the barrier between blood and brain. These reinforced defences protect the brain from many potentially harmful molecules and micro-organisms, but also presents a challenge for the administration of drugs to the brain. (Magnification: unknown)

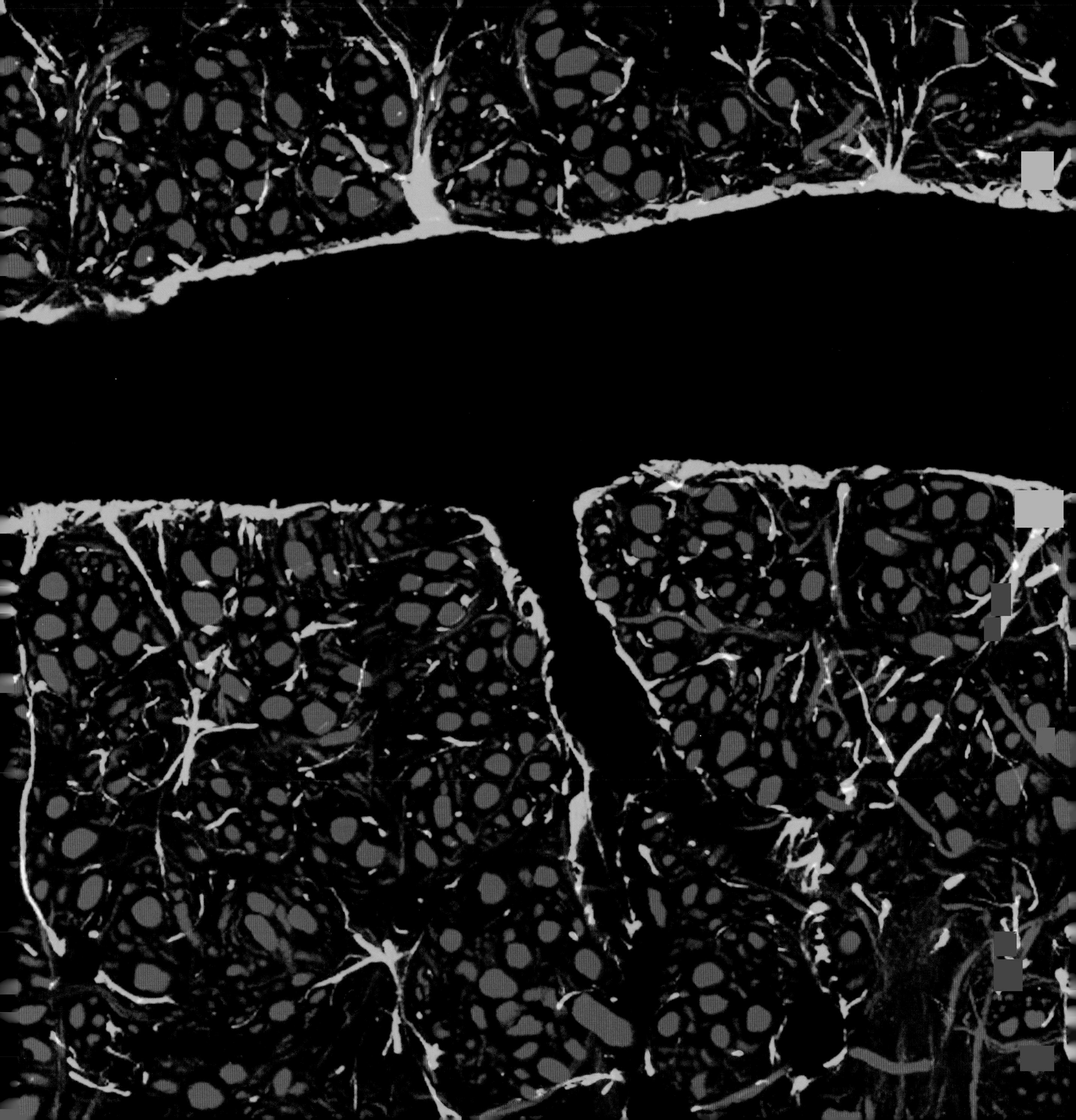

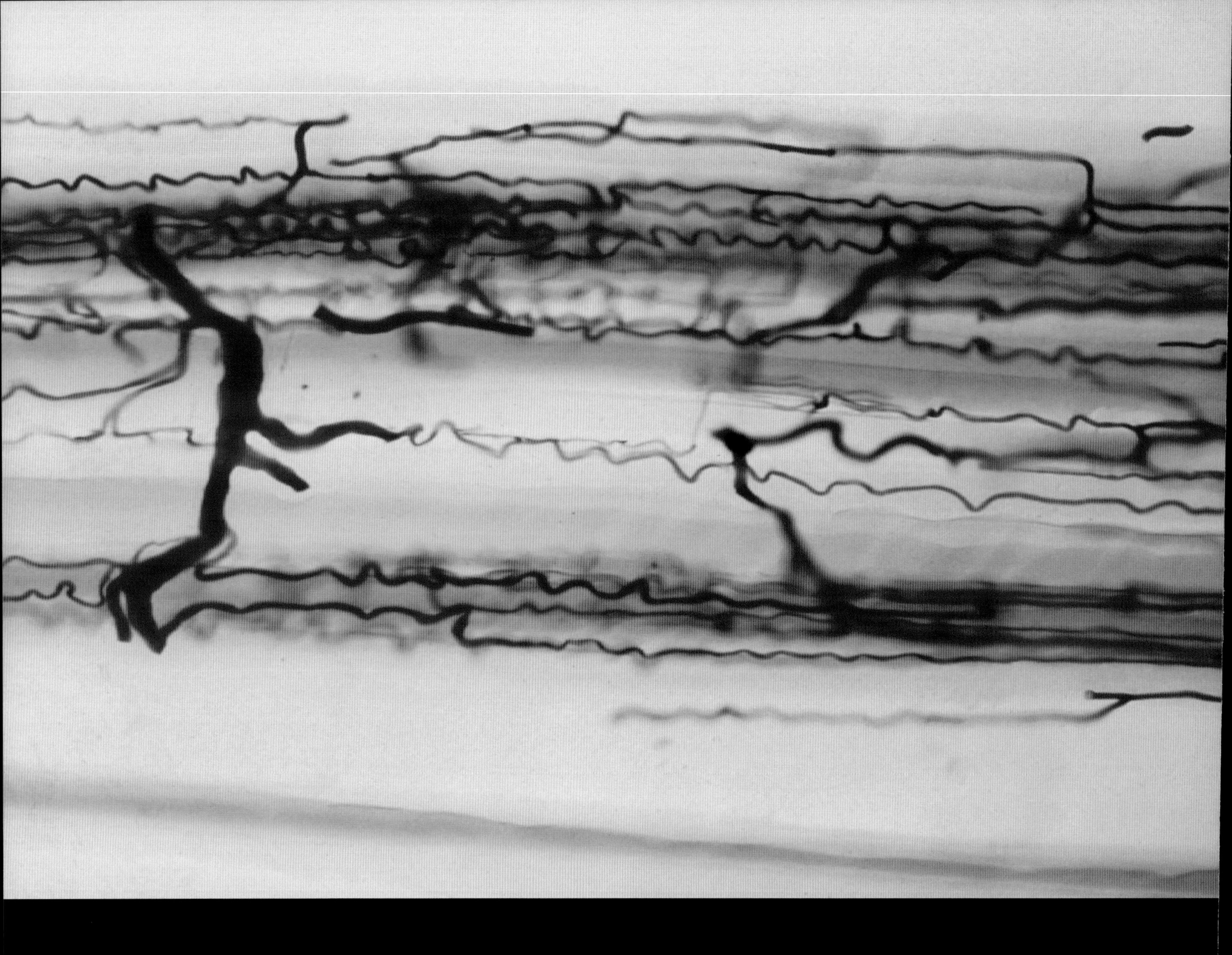

Above: Capillary blood supply in skeletal muscle (light micrograph)
The capillaries are the very fine blood vessels through whose thin walls oxygen and nutrients pass from the blood to the body. Skeletal muscles, as their name implies, move the bones of the skeleton, according to our instructions. (The so-called 'smooth' muscles around our organs, by contrast, require no conscious instruction.) Here, the muscle fibres (pale pink) have been teased apart to reveal the capillary bed (to which red dye has been added to aid identification). (Magnification: x64 at 3.5cm/1¼in size)

Right: Blood clot (coloured scanning electron micrograph)
Red blood cells have been trapped by a web of thin yellow-white strands of fibrin. Fibrin is an insoluble protein produced by platelets (fragments of white blood cells) from a soluble protein called fibrinogen normally present in blood. Blood clots may occur on the surface of skin in case of injuries or inside blood vessels. These internal clots, known as thrombi, may be caused by having too many platelets. They can lead to heart attacks. (Magnification: x1830 at 6x7cm/2¼x2¾in size)

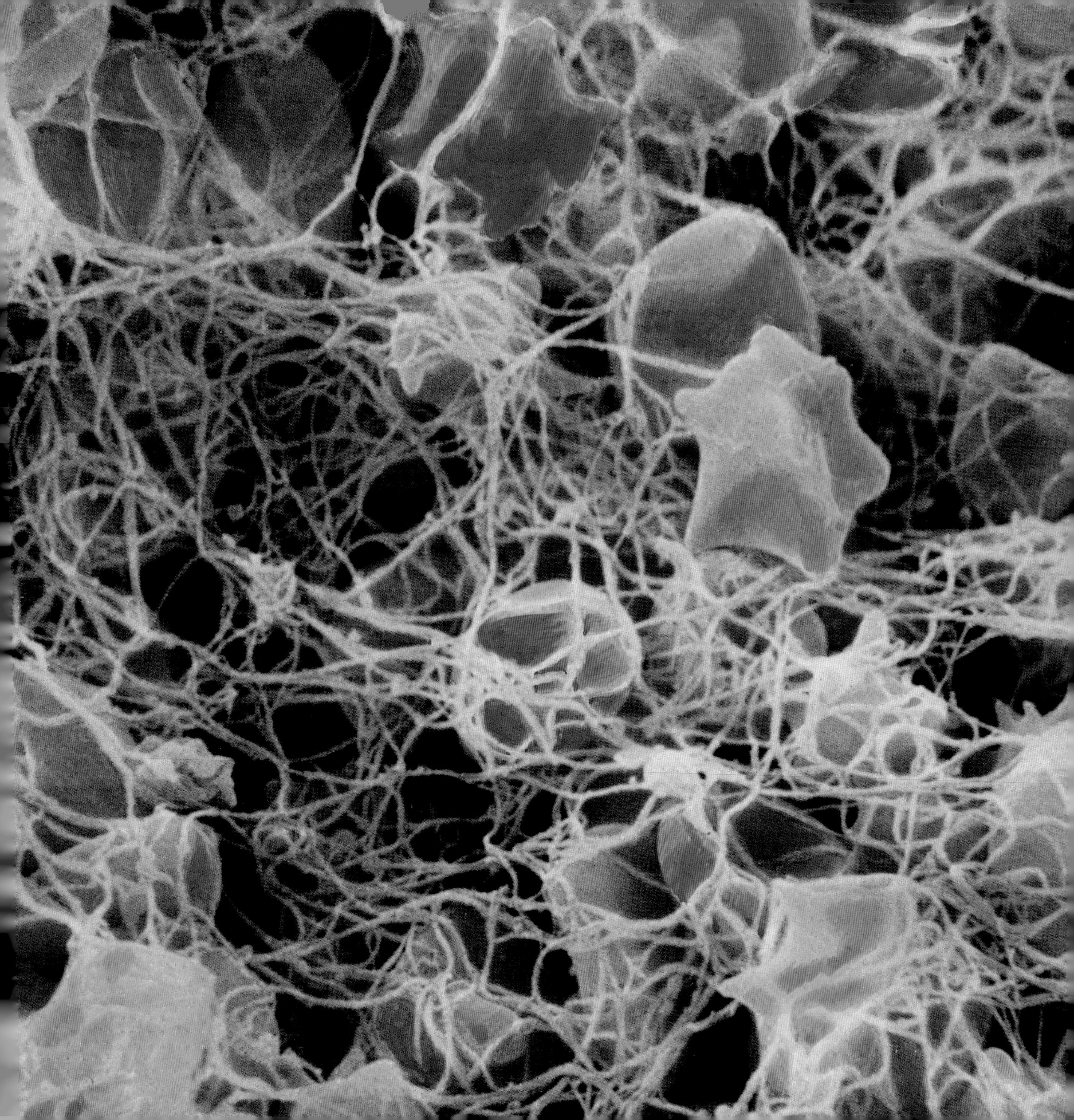

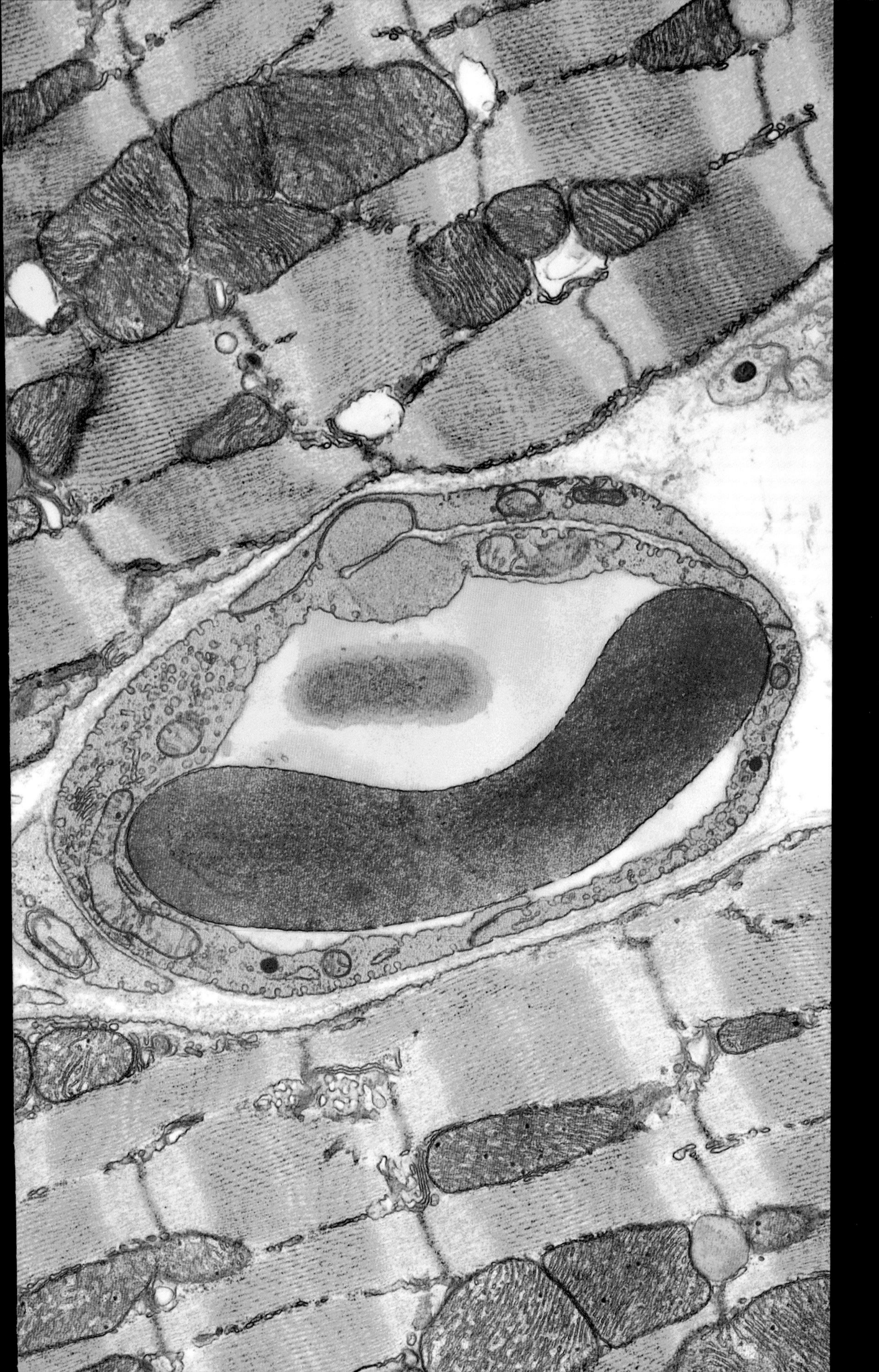

Far left: Mast cell (coloured transmission electron micrograph)
A mast cell is a white blood cell which provides the immune system's response to inflammation. The central oval structure (light green) is the cell's nucleus. It is surrounded by a body (the cytoplasm, dark green) which carries granules (in pink). The granules contain serotonin, heparin (an anti-coagulant) and histamine. Histamine increases the permeability of blood vessel walls, allowing healing white cells to pass through. The granules are released in response to inflammation or allergy. (Magnification: x8750 at 6x7cm/2¼x2¾in size)

Left: Cardiac muscle (coloured transmission electron micrograph)
Cardiac muscle (green) is under subconscious control and continuously contracts to pump blood around the body without tiring, unlike skeletal muscle. This is because of the presence of mitochondria (purple) which supply the muscle cells with energy. The muscle fibre is divided by fine walls (dark green) into individual contraction cells called sarcomeres. The central object is a capillary (orange) seen in cross-section, with a red blood cell passing through it. (Magnification: x3600 at 10cm/4in size)

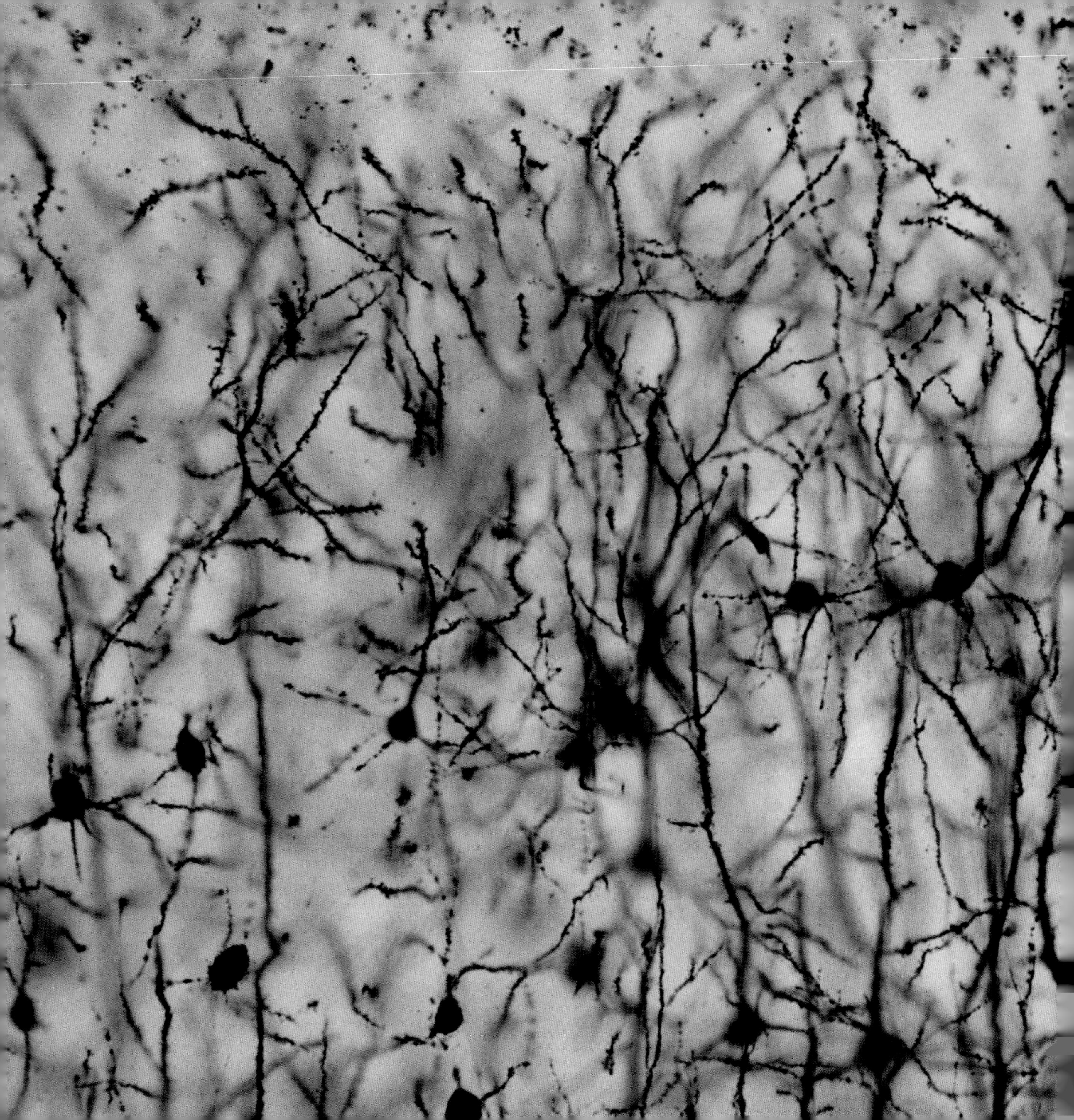

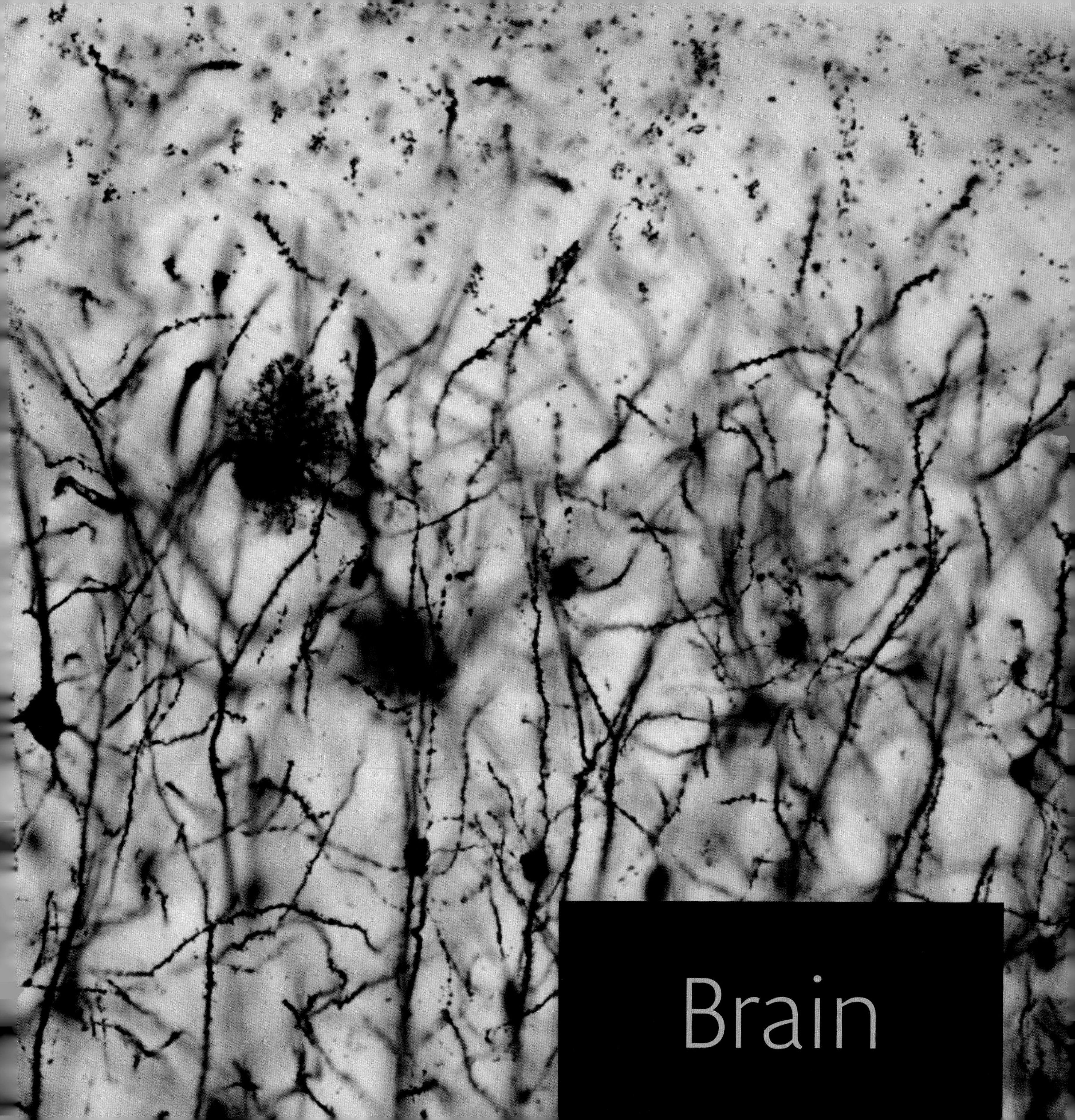
Brain

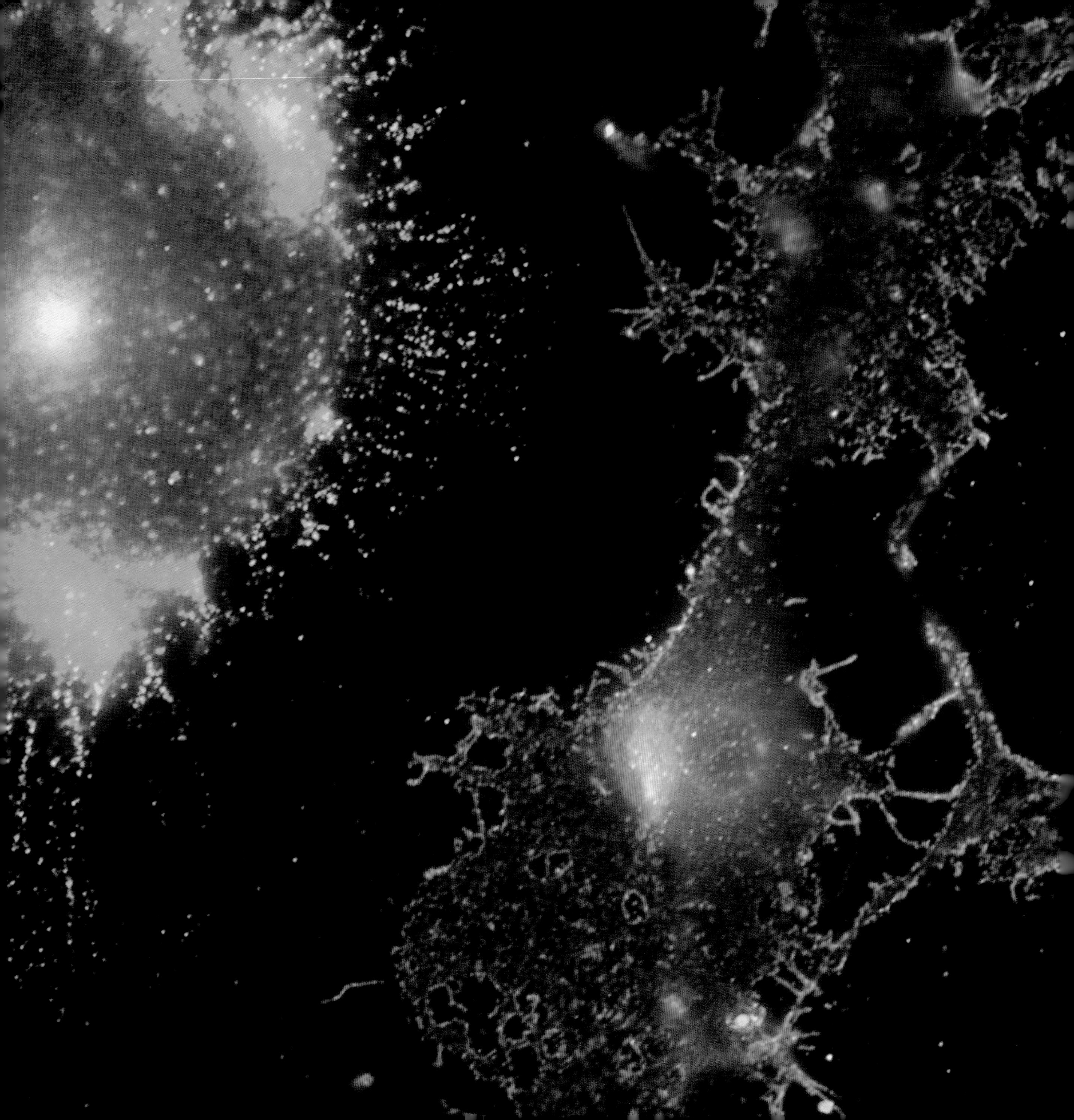

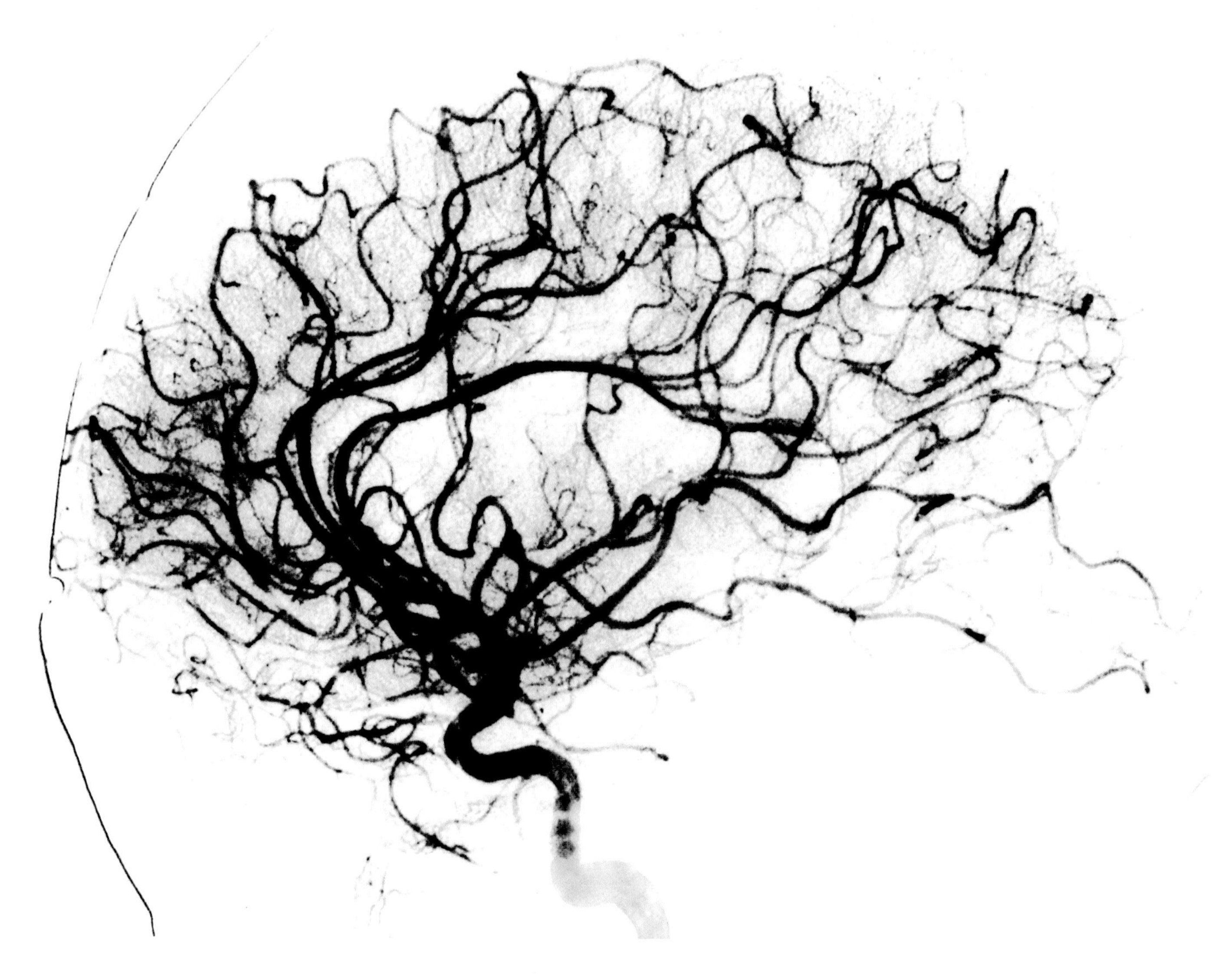

Left: Brain cells in culture (fluorescence light micrograph)
This image shows two important support cells (glial cells) of the human brain. The green splash is a microglial cell, which responds to immune reactions in the central nervous system. Microglial cells recognize areas of damage and inflammation and swallow cellular debris. The larger orange shape is an oligodendrocyte. The ragged extensions of an oligodendrocyte can supply many neurons (nerve cells) with myelin, an insulating material which allows each neuron's communicating axon to transmit electrical impulses efficiently. (Magnification: unknown)

Above: The brain's blood supply (cerebral angiogram)
This striking side-view of a healthy brain, facing to the left, was achieved by injecting trace dye into the internal carotid artery (entering the brain from the bottom of the picture). That artery supplies the brain with vital oxygen-bearing blood. Different sections of the artery supply different parts. In the image you can clearly see the thick, dark horizontal section where the artery enters the brain, called the inferolateral trunk, from which smaller arteries supply the jaw, mouth and nose. (Magnification: unknown)

Previous page: Cerebral cortex nerve cells (light micrograph)
Neurons (nerve cells) allow information to be received, interpreted and relayed around the body. These ones are in the grey matter of the cerebral cortex, which forms the outer layer of the brain responsible for conscious thought, memory and language. From each cell, branching extensions (dendrites) receive impulses from other neurons and pass them to the cell body. Each neuron also has an extension called an axon through which it relays impulses to the dendrites of other neurons. (Magnification: unknown)

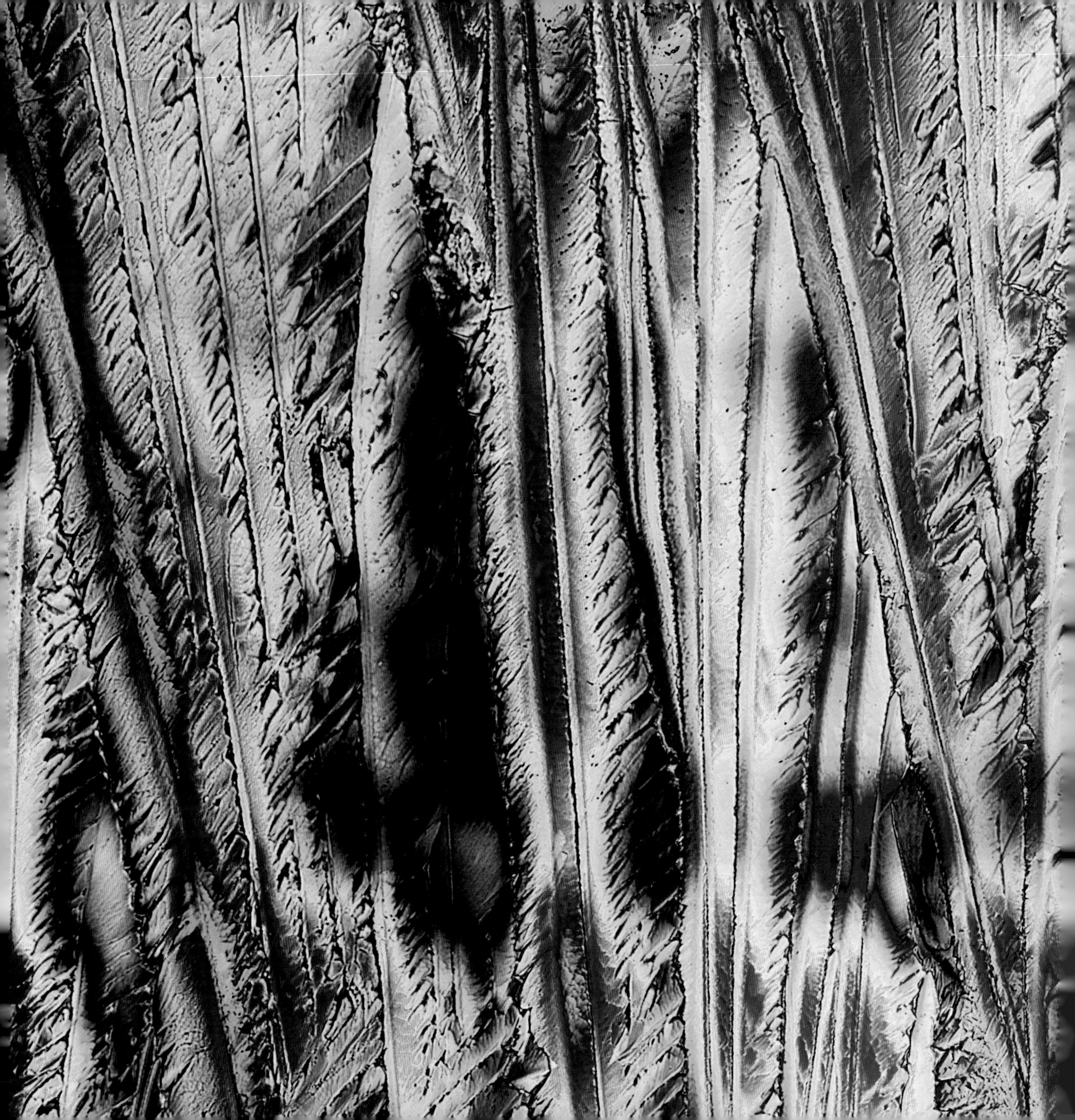

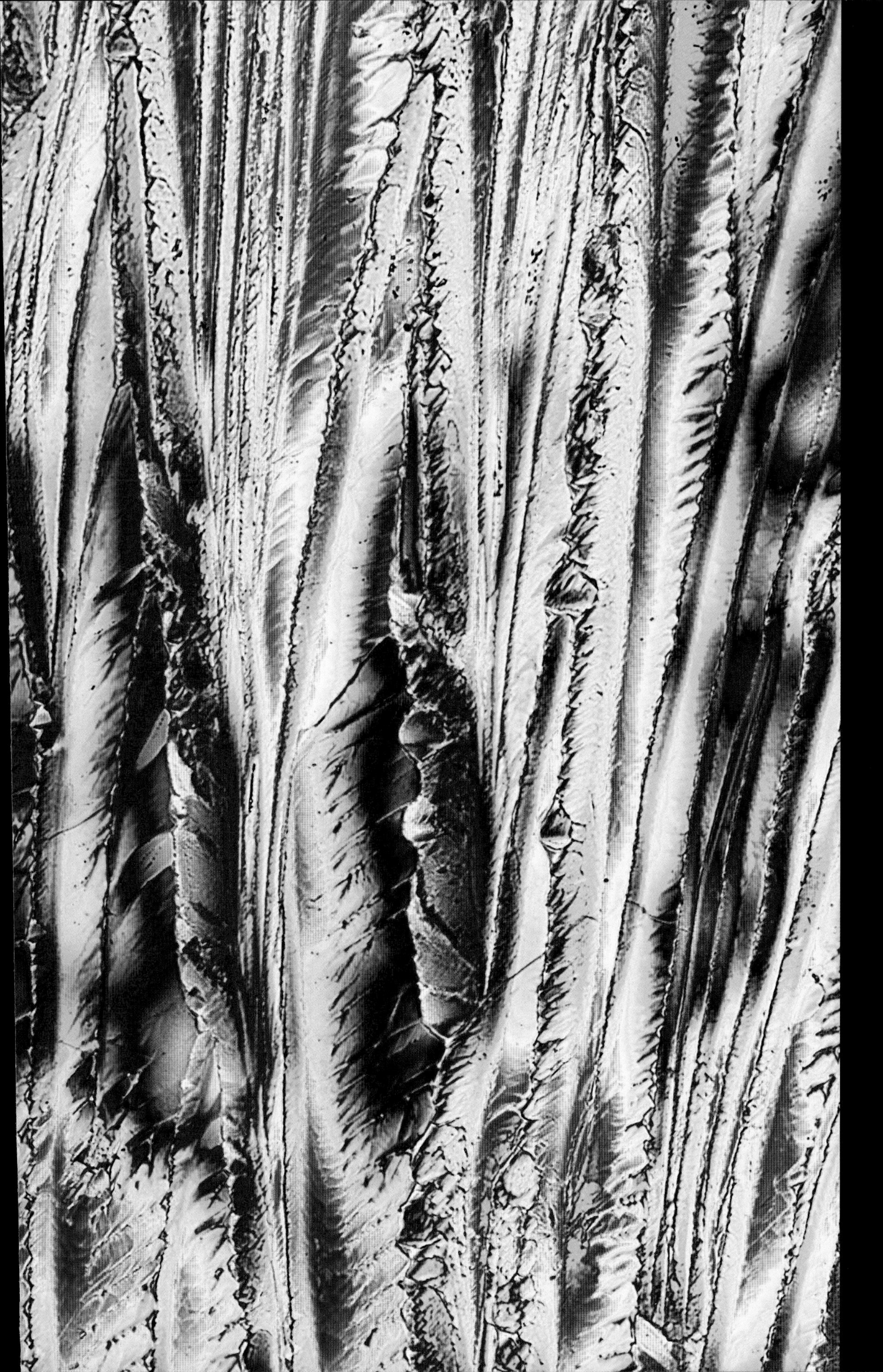

Dopamine neurotransmitter (polarized light micrograph)

Dopamine is a chemical transmitted by neurons in the nervous system to neighbourin neurons or to muscle cells. It is released when a pleasurable new stimulus is encountered, and directs the brain to find more of the stimulus. It therefore plays a significant role in addiction, not only to drugs that release or mimic dopamine (such as cocaine) but also to social addictions such as gambling. Lack of dopamine is linked to several disorders, including Parkinson's disease, depression and schizophrenia. (Magnification: unknown)

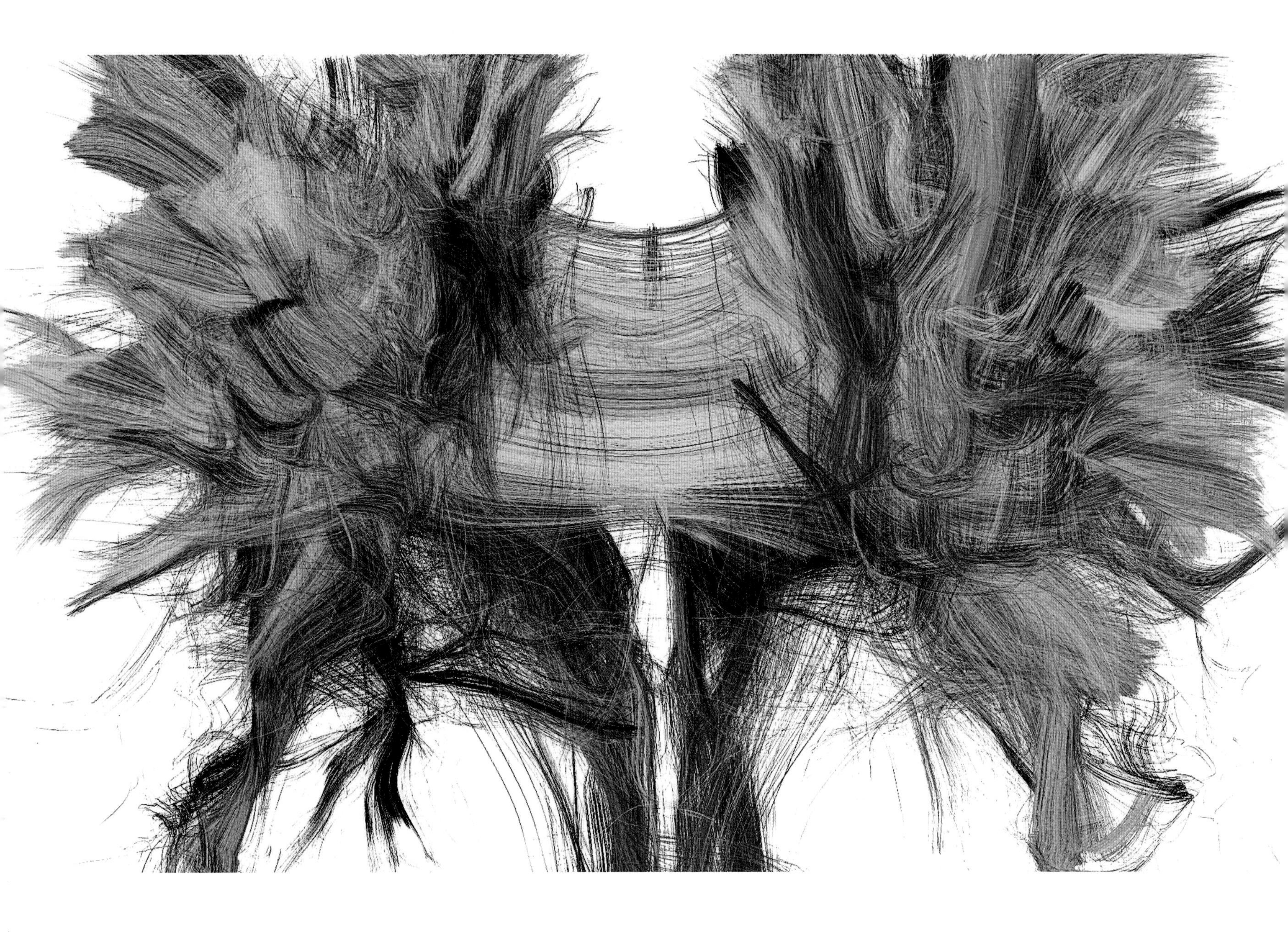

Above: Corpus callosum (magnetic resonance imaging scan)
A technique called coronal 3D diffusion tensor imaging measures the direction of water diffusion, which in the brain reveals the orientation of nerve fibres. The technique is also known as tractography. The resulting image, known as a tractogram, reveals these lively brush strokes, actually nerve pathways (blue) in and around the brain's corpus callosum. The corpus callosum (centre) is the band of nerve fibres that connects the left and right hemispheres of the brain. (Magnification: unknown)

Right: White matter fibres (diffusion spectrum imaging MRI scan)
In the brain, neurons (nerve cells) occupy the outer layer, called the grey matter. The white matter within consists of axons, the long extensions of each nerve cell which transmit nerve signals between brain regions and between the brain and the spinal cord. This is a scan of the bundles of white matter nerve fibres, made by diffusion spectrum imaging in which a magnetic field reveals the water contained in the fibres, thus mapping their criss-crossing patterns. (Magnification: unknown)

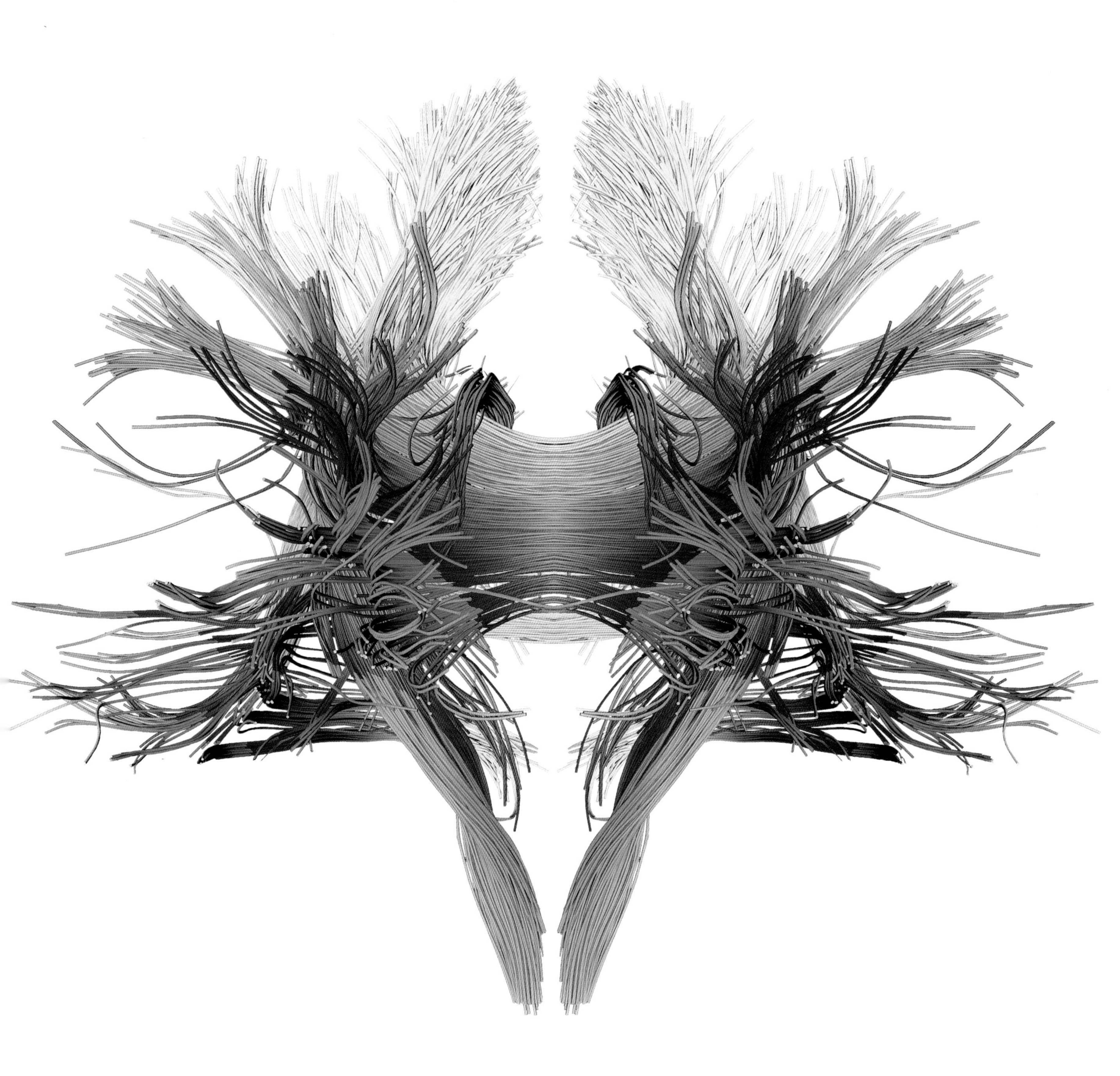

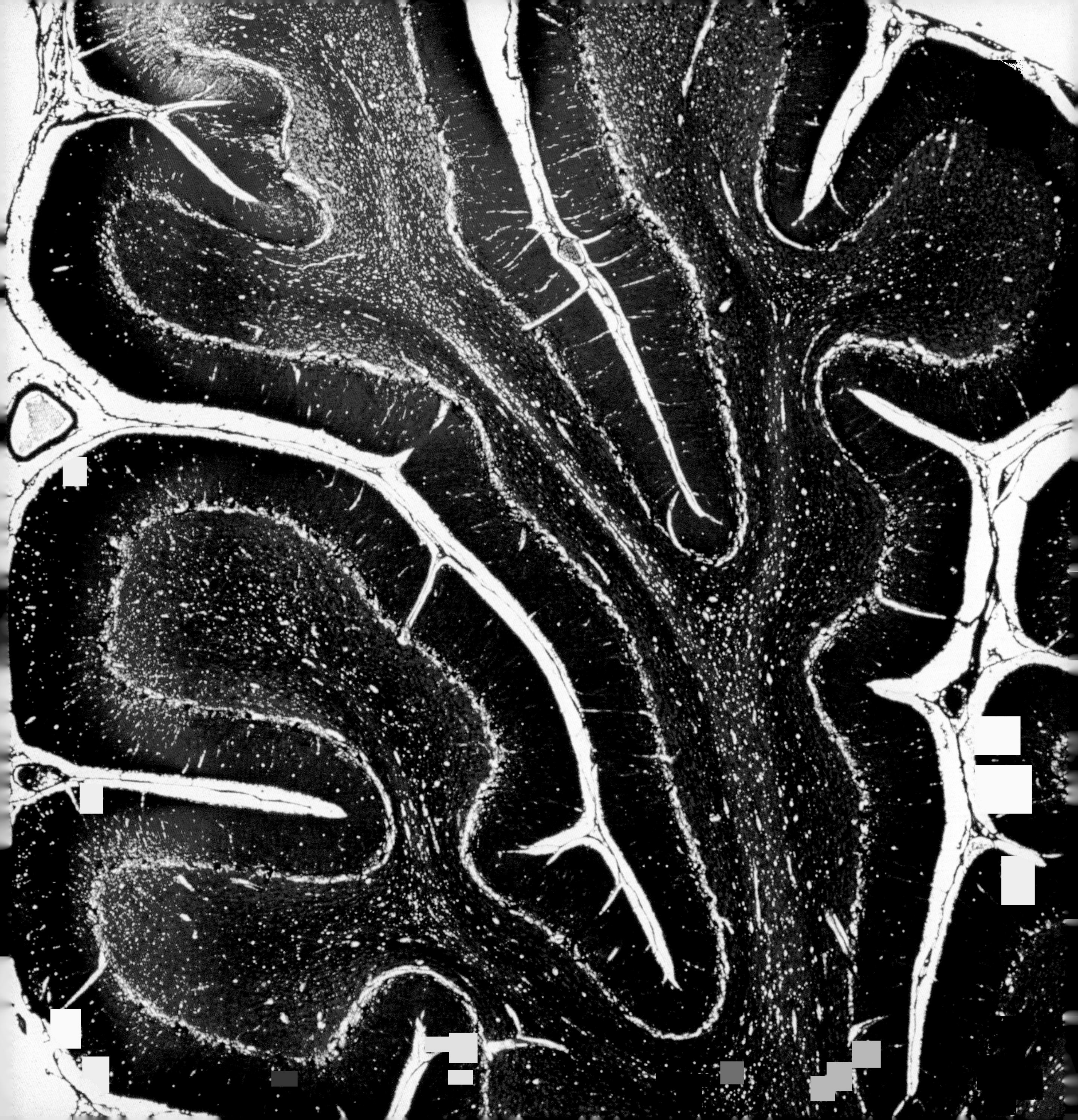

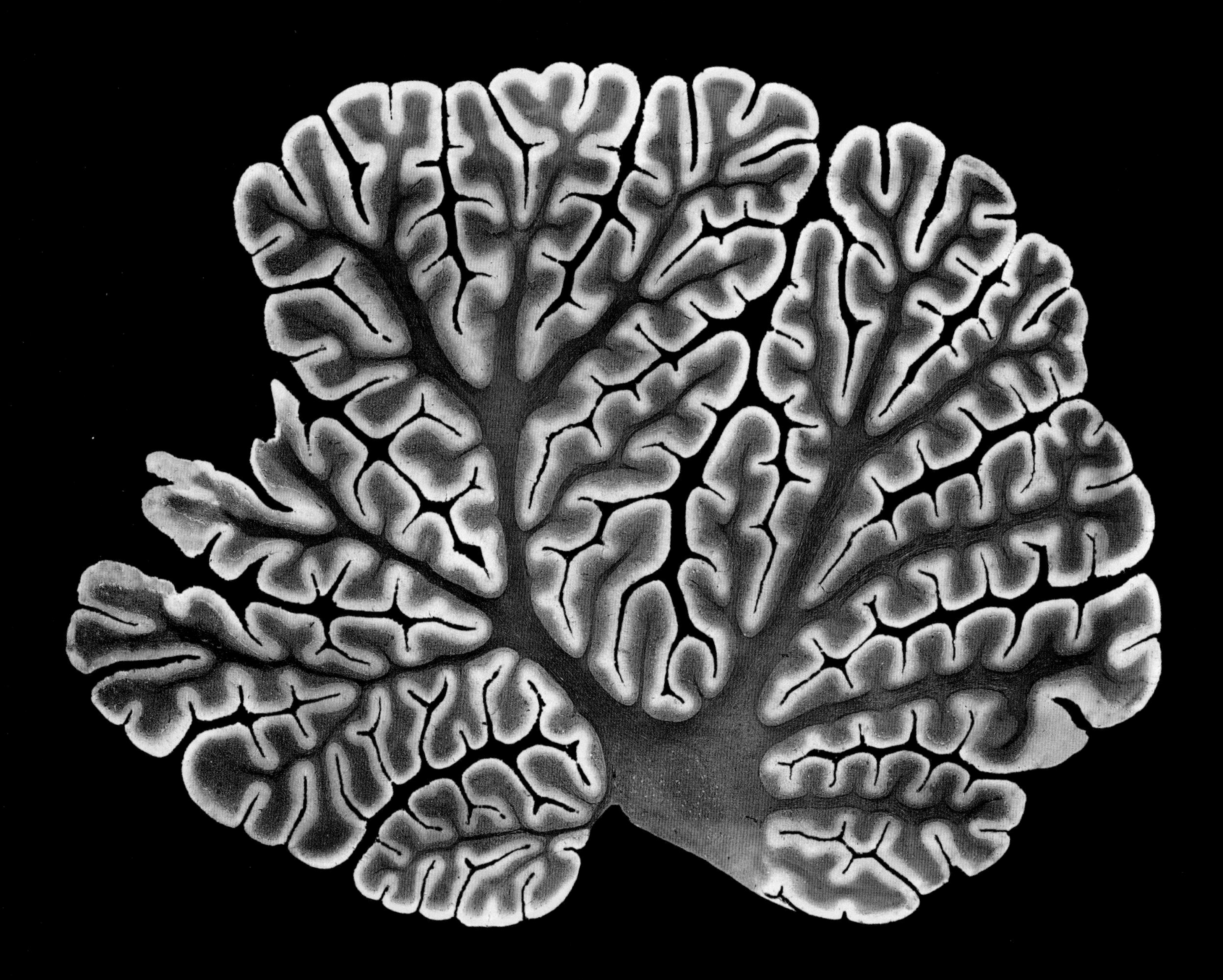

Left: Section through the cerebellum (light micrograph)
The cerebellum controls movement, posture and balance. It is a large, highly folded structure at the base and back of the brain. In this micrograph, cells in the outer portion of each fold are stained more darkly. Unusual neurons, called Purkinje cells, inhabit the thin white band between the inner and outer layers. They are responsible for coordinating movement, and are vulnerable to toxins such as alcohol and lithium. Densely packed nerve fibres form the central core of each fold. (Magnification: x30 at 6x7cm/2¼x2¾in size)

Above: Cerebellum structure (coloured light micrograph)
The cerebellum (literally 'small brain') sits below and behind the cerebrum or 'main brain'. It controls movement and has a role in attention span and language. Although only 10% of the brain by volume, it has around 70% of the brain's neurons. Each neuron in the outer molecular layer (shown yellow here) 'speaks' to other neurons through an extension called an axon. Axons form the dense inner branches of the cerebellum (here dark orange). (Magnification: x8 at 10cm/4in size)

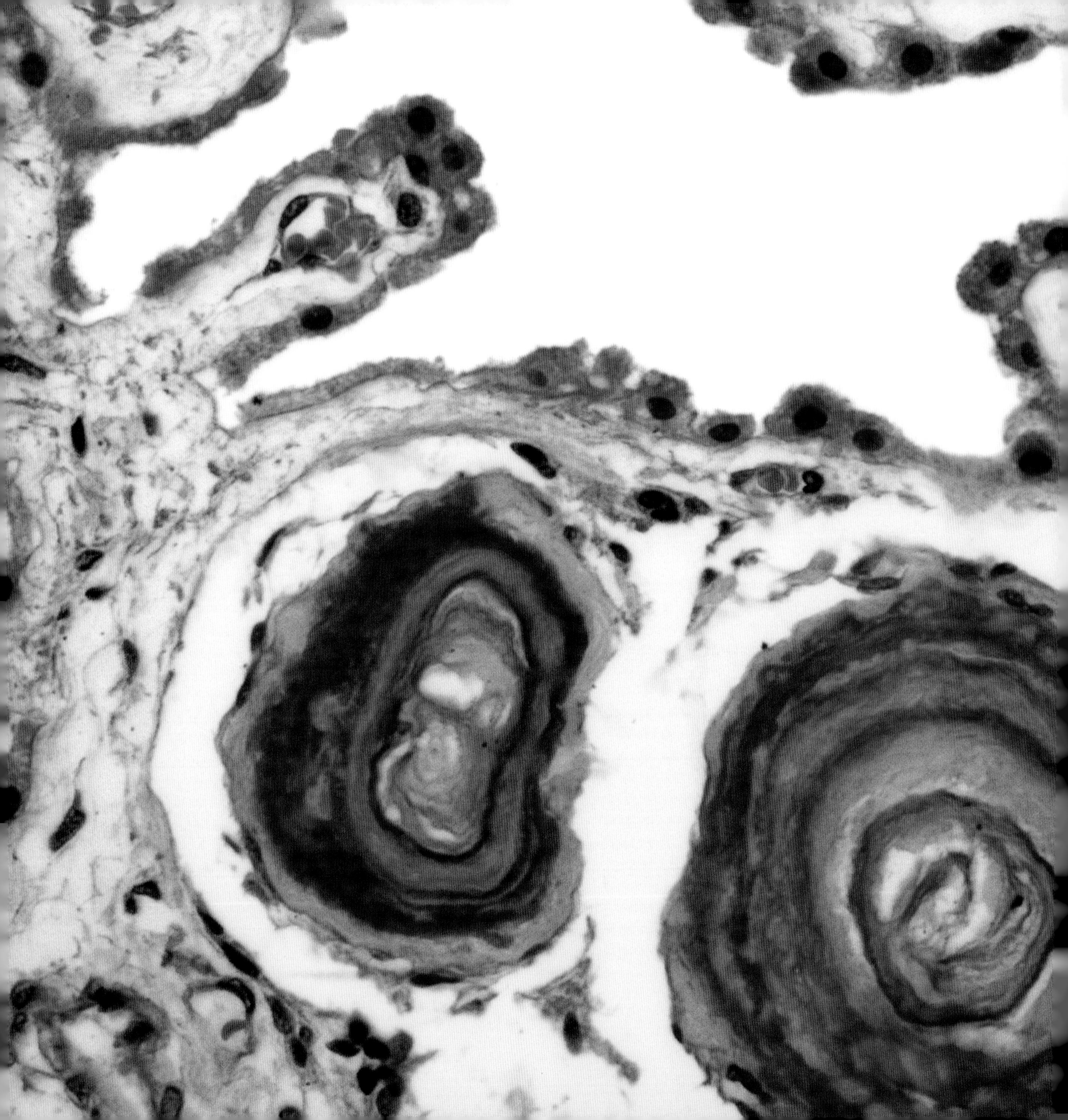

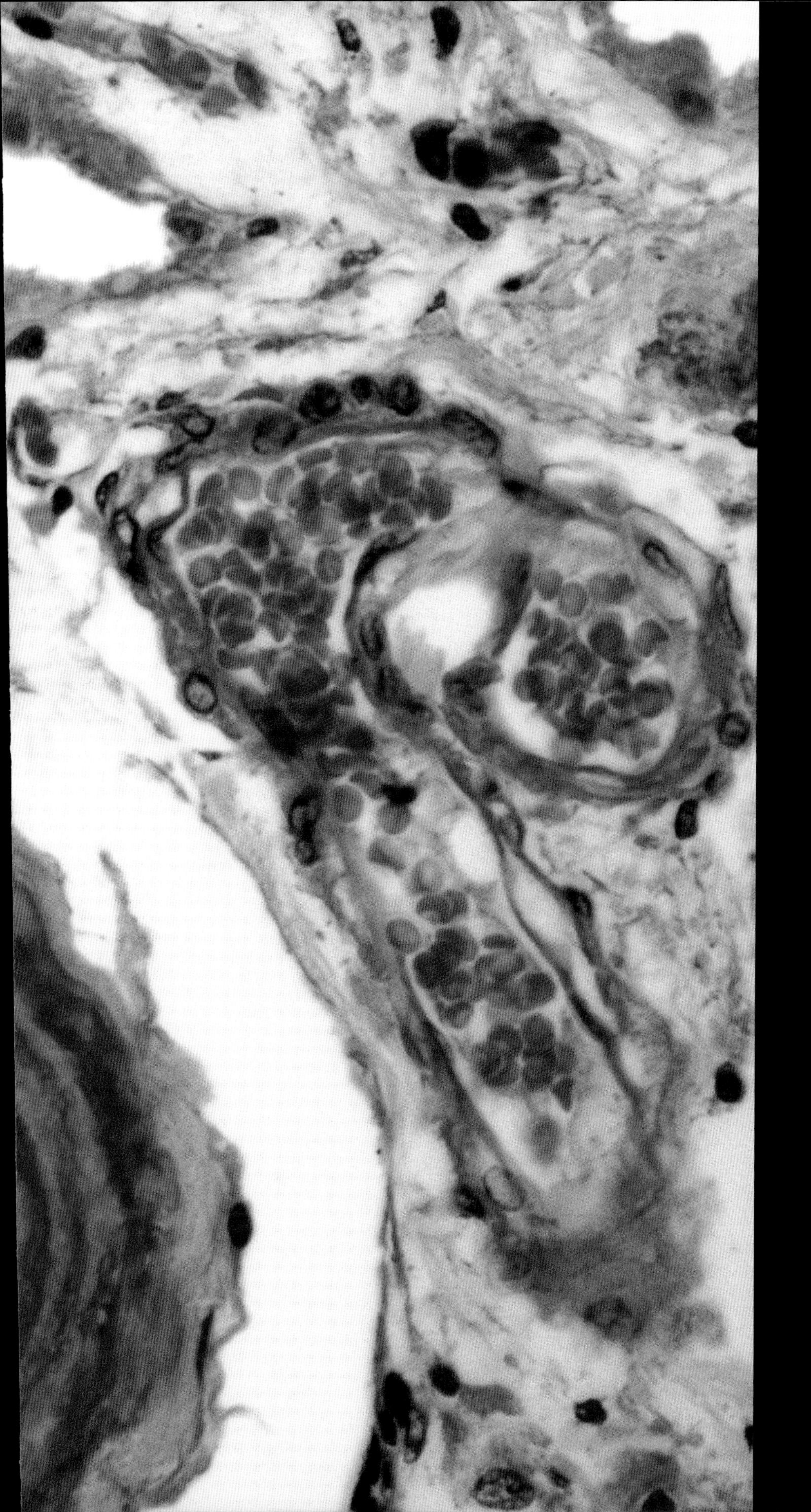

Corpora amylacea (light micrograph)

Corpora amylacea are small lumps found generally in aging hearts, prostate glands or brains. They are normal, and usually found in healthy organs, but their function remains unknown. They are formed from redundant cells or thickened secretions of fluid. The two in this example (purple, bottom) are in an area of capillaries in the brain, the choroid plexus, which produces a protective liquid cushion for the brain called cerebrospinal fluid. (Magnification: x200 at 10cm/4in size)

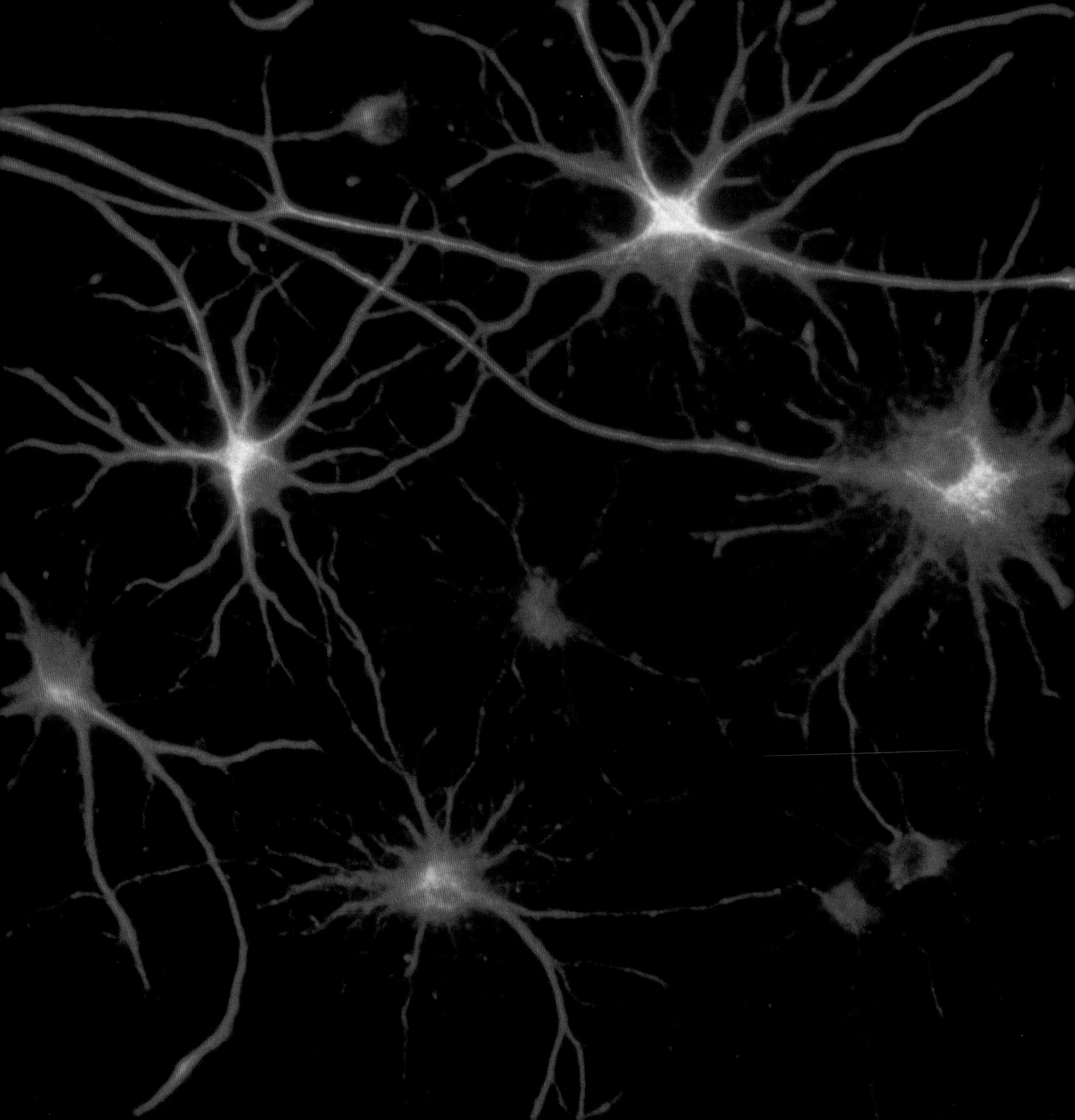

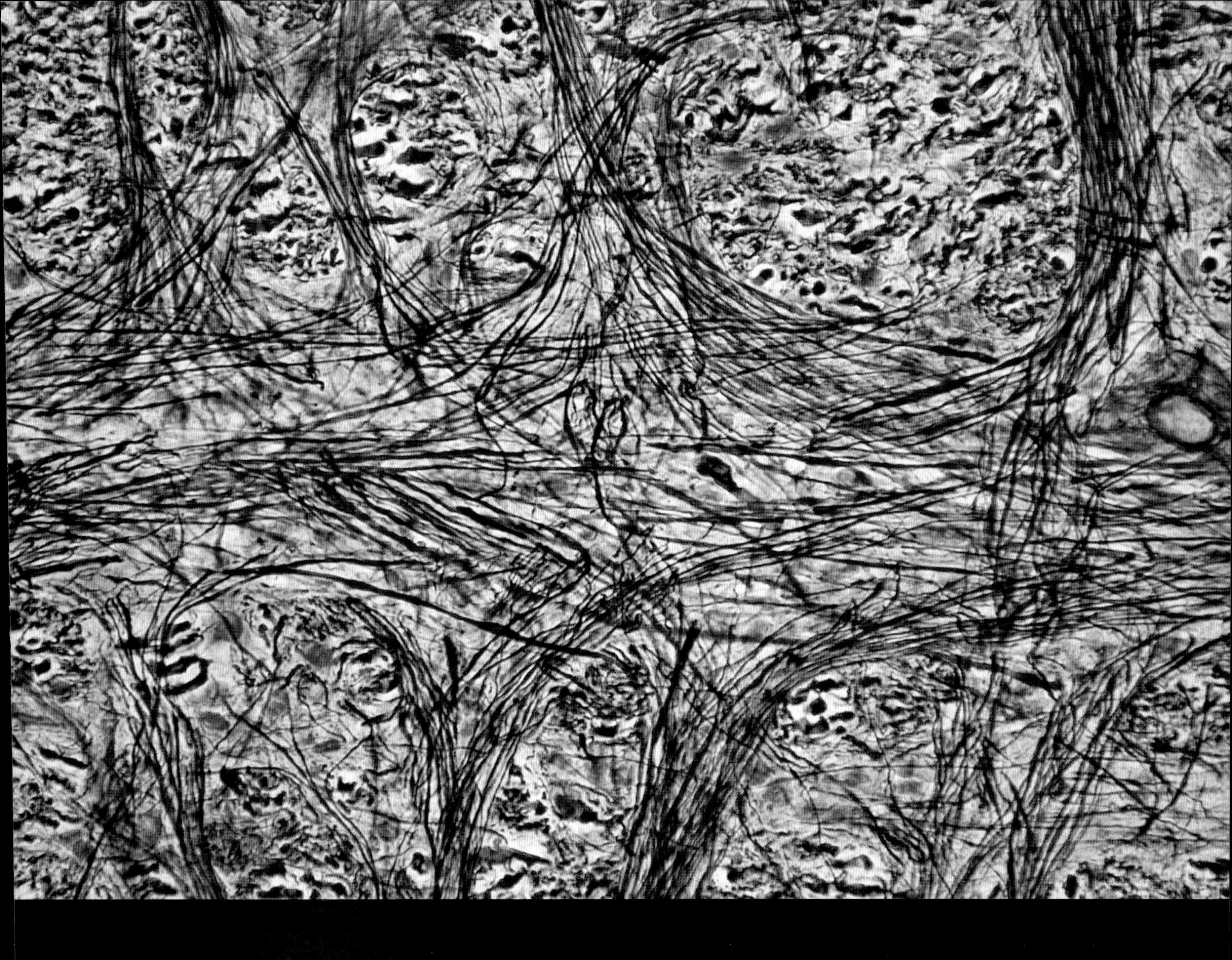

Left: Astrocytes from human brain tissue (fluorescence light micrograph)
Astrocytes form part of the neuroglia ('nerve glue'), the mass of support cells found in the grey and white matter of the brain and spinal cord. They are the most numerous such cells in grey matter, where they are believed to provide mechanical and metabolic support for neurons (nerve cells). They also have a role in the regeneration of neurons following injury to the brain or spinal cord. Here their nuclei appear yellow, and their dendrites (tendrils) red. (Magnification: x120 at 5x7cm/2x2¾in size)

Above: Nervous tissue of the brain (light micrograph)
All nerve cells, or neurons, consist of a cell body with a varying number of extensions called the cell's 'processes'. One cell body is visible at far right, containing a central, circular nucleus. Processes of two types extend from the cell body: a single axon and one or more dendrites. Axons are the channels for communicating information to other neurons. Here, they have been chemically stained to show up as dark lines, running horizontally and vertically over the image. (Magnification: x250 at 6x7cm/2¼x2¾in size)

Above: Brain pathways (coloured magnetic resonance imaging scan)
A view of the white matter pathways of the brain – white matter is composed of nerve cell fibres that carry information between nerve cells in the cerebrum of the brain (top half of image) and the brain stem (bottom centre). Blue represents neural pathways from the top to the bottom of the brain, green represents pathways from the front (left) to the back (right), and red shows pathways between the right and left hemispheres of the brain. (Magnification: unknown)

Right: Astrocyte brain cells (immunofluorescence light micrograph)
Astrocytes have numerous branches of connective tissue which work to nourish and repair neurons (nerve cells). They may also play a role in information storage. Immunofluorescence is a staining technique which uses antibodies to attach fluorescent dyes to specific tissues and molecules in the cell. In this image the cytoplasm of astrocytes (their extended star-shaped bodies) has been stained green, and their nuclei blue. The blue nuclei of other cells are also seen. (Magnification: x125 at 6x4.5cm/2¼x1¾in size)

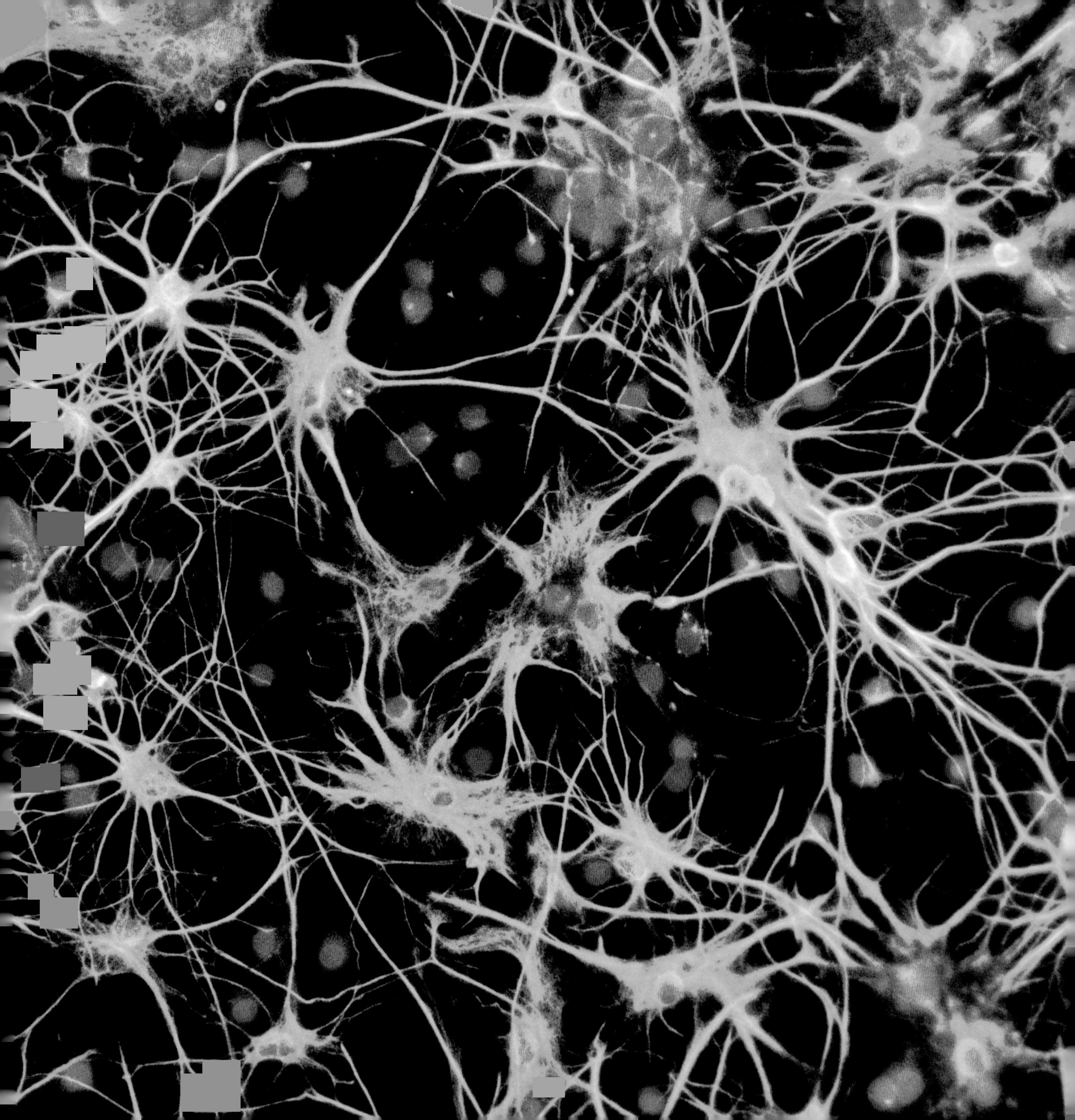

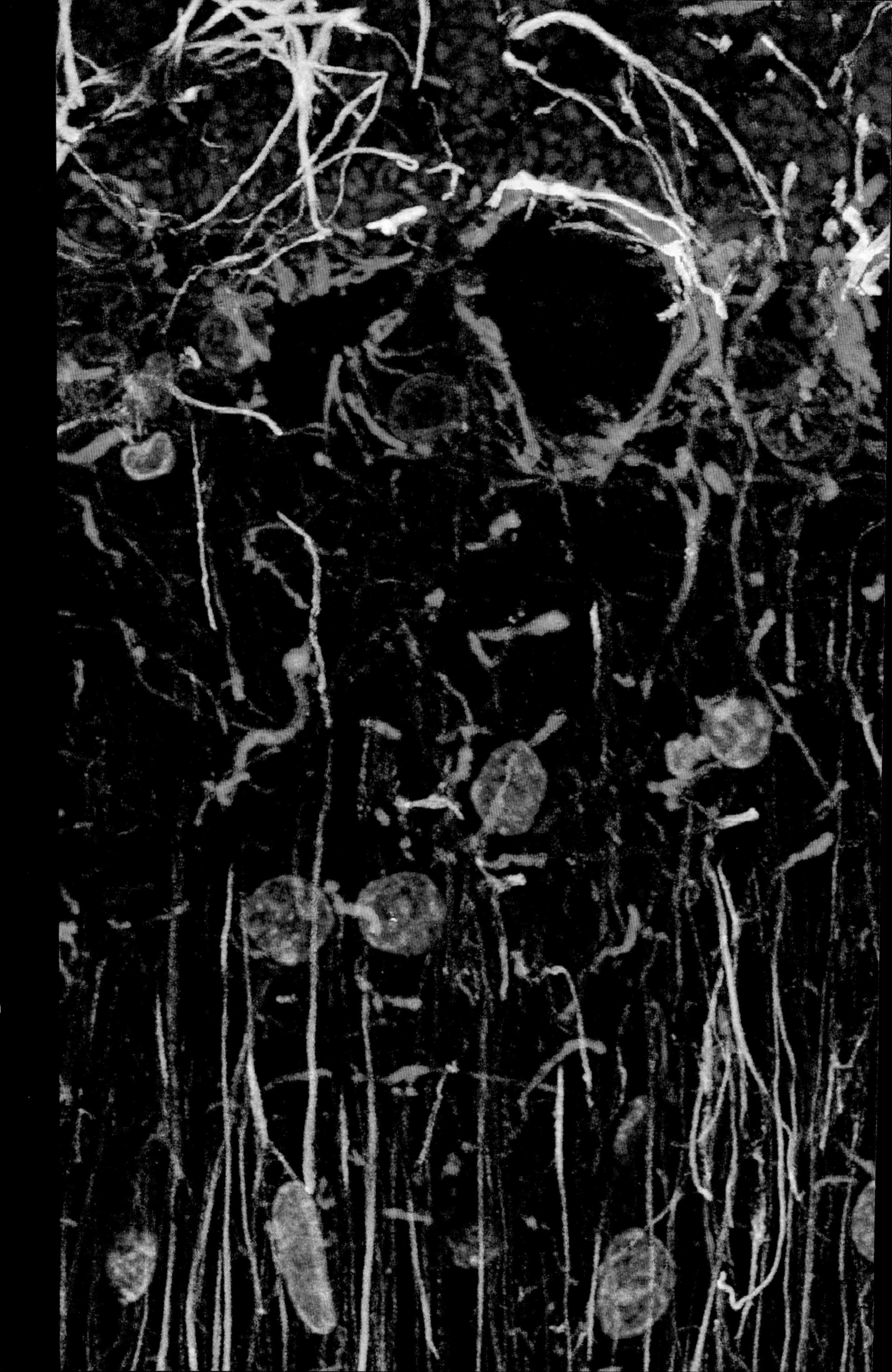

Cerebellum tissue (coloured confocal light micrograph)

The cerebellum controls balance, posture and muscle coordination. Purkinje cells, a type of neuron (nerve cell), play a crucial role in that function. They form the connection between two layers of the grey matter of the cerebellum. Shown here in red, Purkinje cells have a flask-shaped cell body with many branches (dendrites) receiving impulses from other cells. Green glial cells provide structural support, nutrients and oxygen for the Purkinje cells; cell nuclei appear blue. (Magnification: unknown)

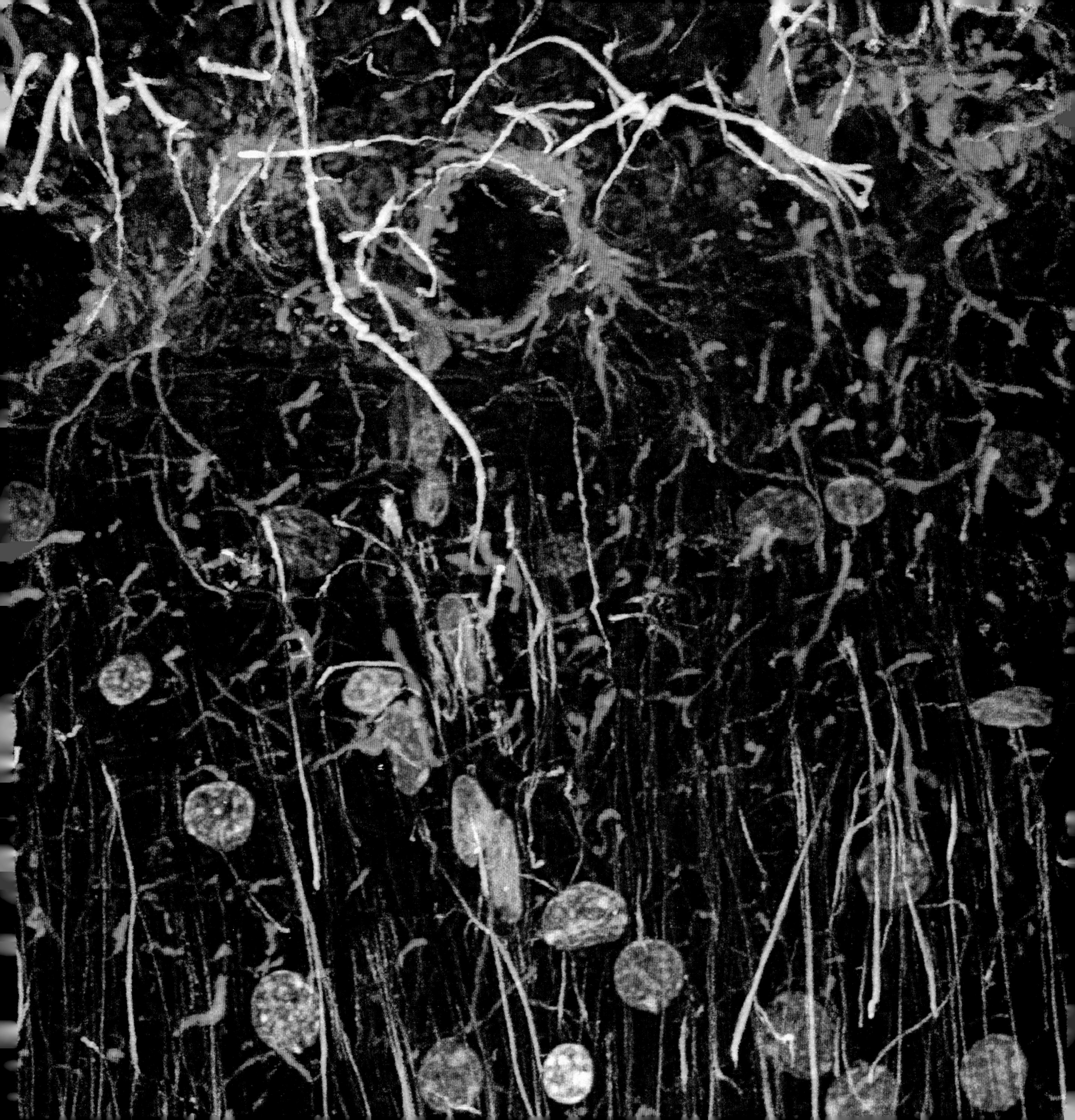

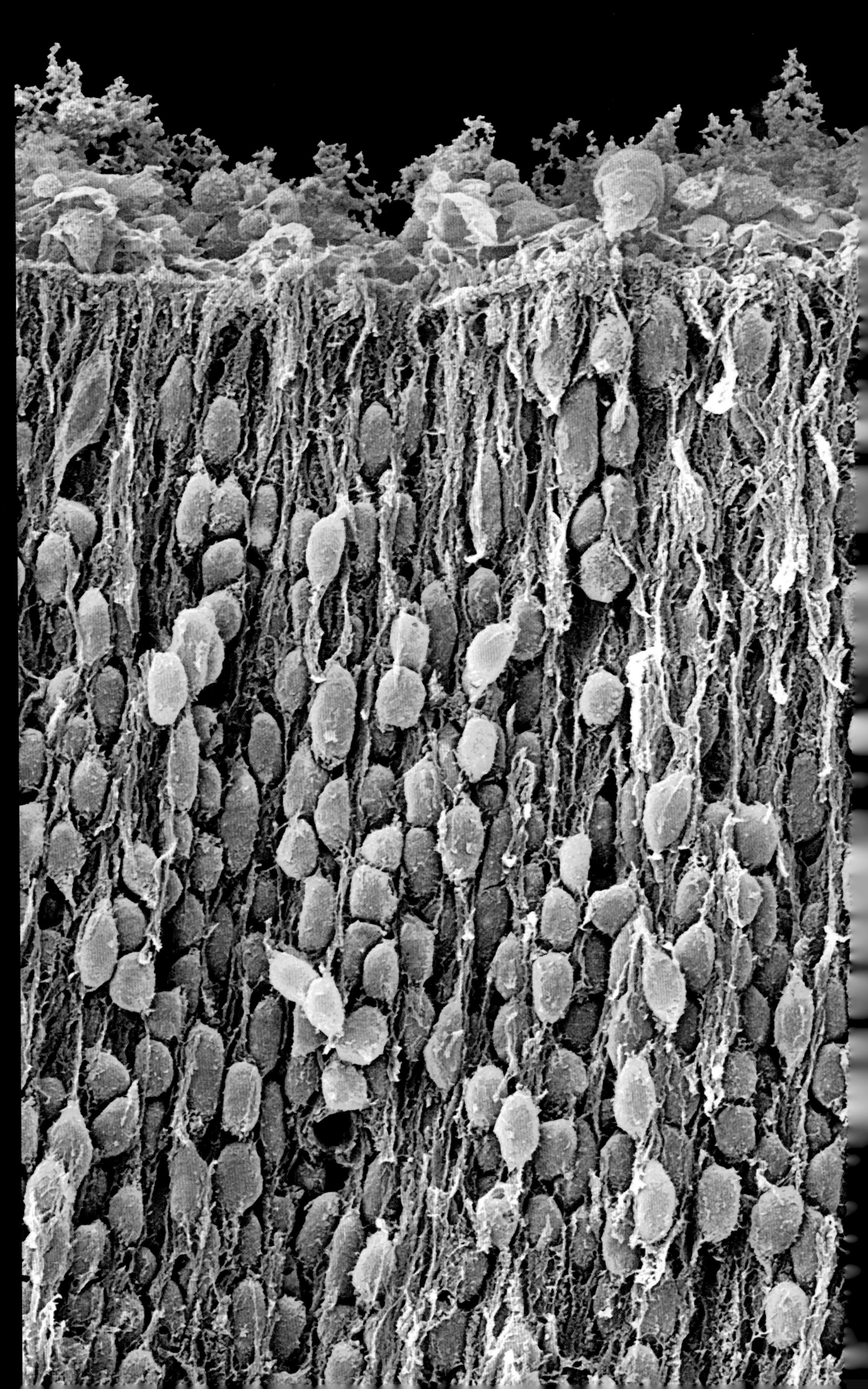

Foetal brain cells (coloured scanning electron micrograph)
The brain begins to develop rapidly in the foetus from around five weeks. This image shows neurons (nerve cells) in the foetal brain, in the area which controls posture and movement. Neurons are responsible for passing information around the central nervous system. Each neuron consists of a cell body (shown yellow here) surrounded by many extensions called dendrites. Here dendrites collect information from other neurons or from sensory cells such as those in the eye and the inner ear. (Magnification: unknown)

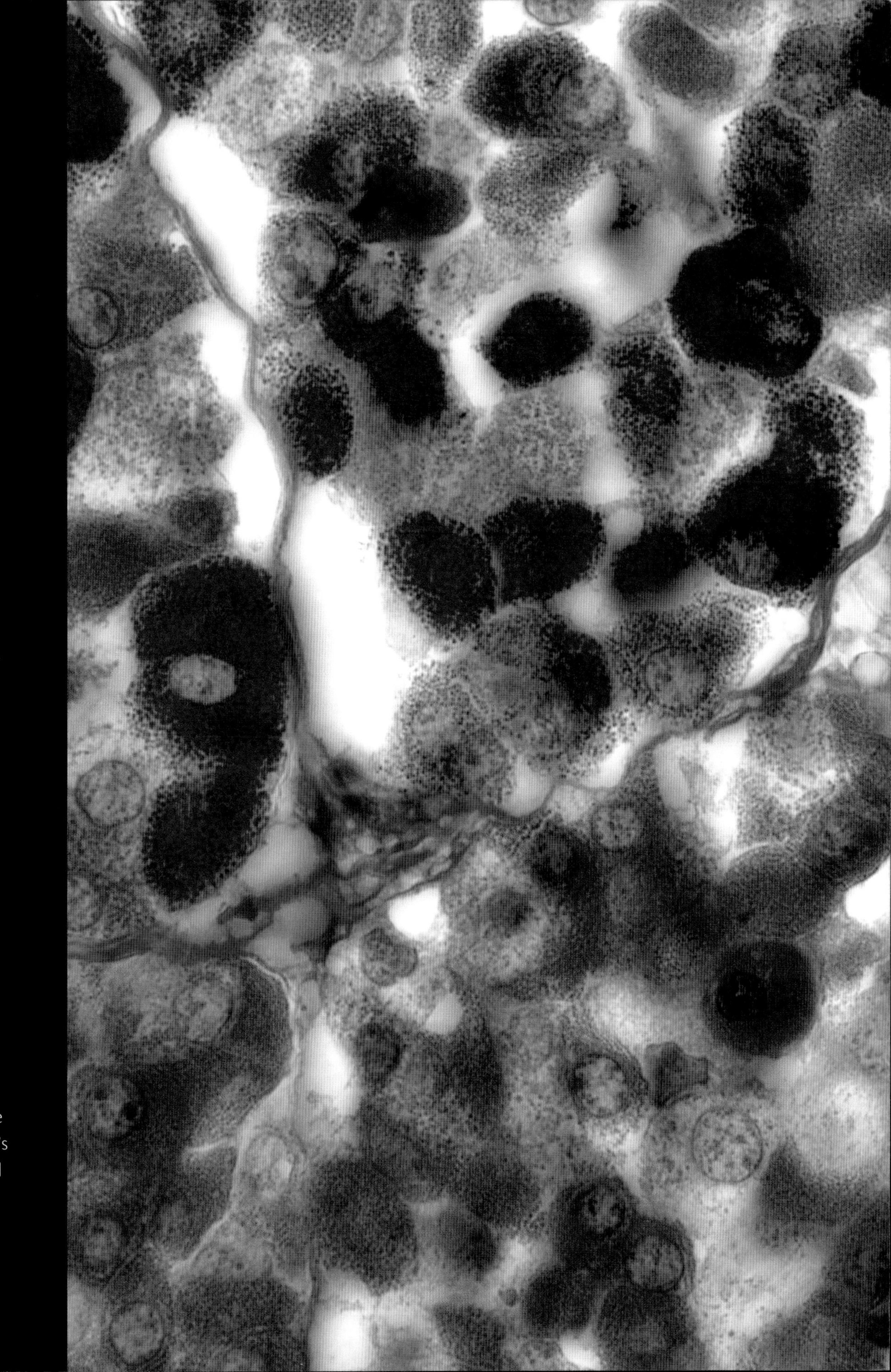

Pituitary gland (light micrograph)

The pituitary gland is one of a network of glands called the endocrine system which secrete hormones directly into the blood. It sits at the base of the brain and regulates aspects of the body's internal environment such as temperature and acidity. The hormones are produced by cells called chromophobes (pale blue). Two kinds of white blood cell are also present as part of the body's immune system: eosinophils (stained red) and basophils (purple) target parasitic worms and allergies. (Magnification: x1100 actual size)

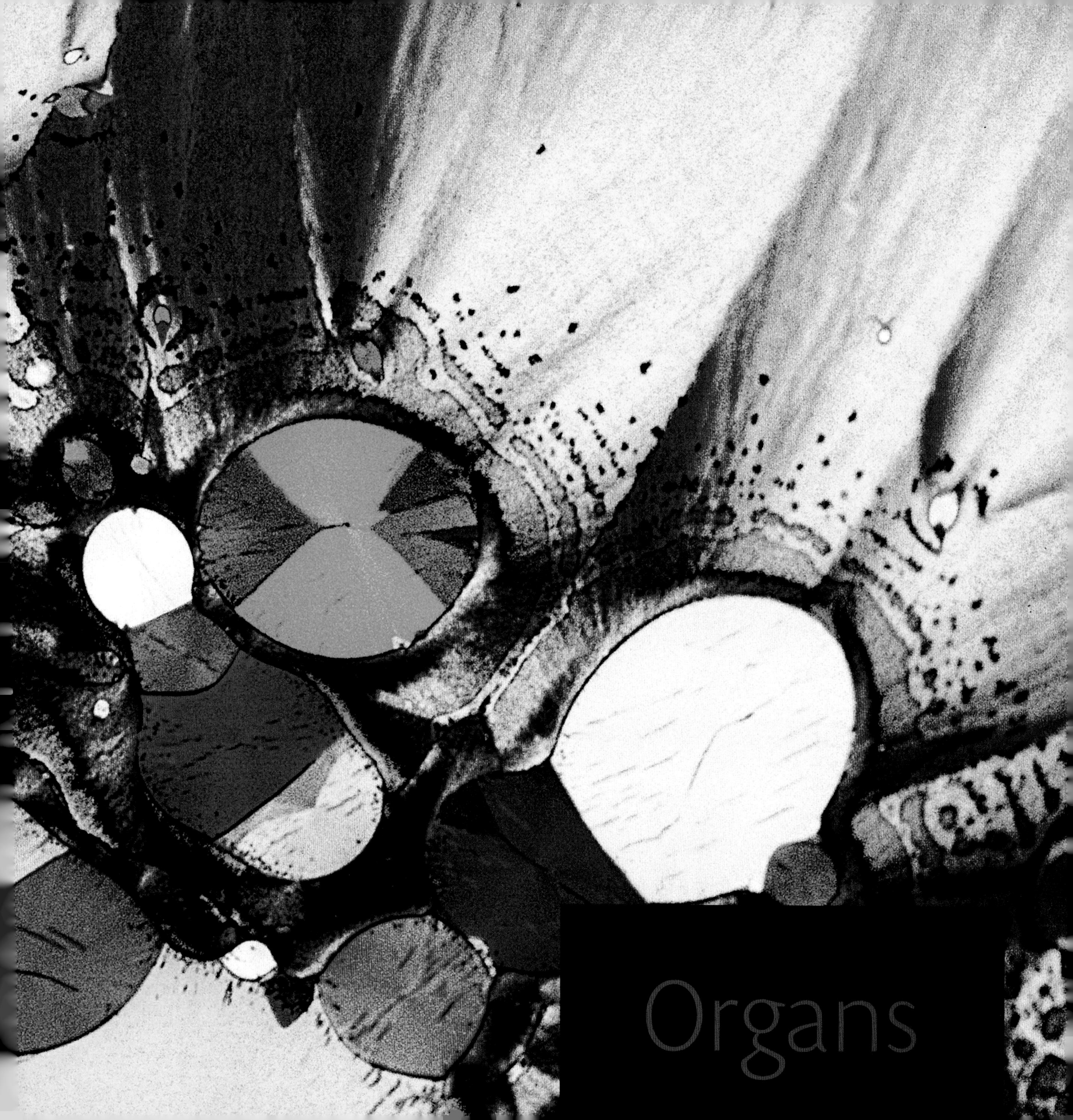
Organs

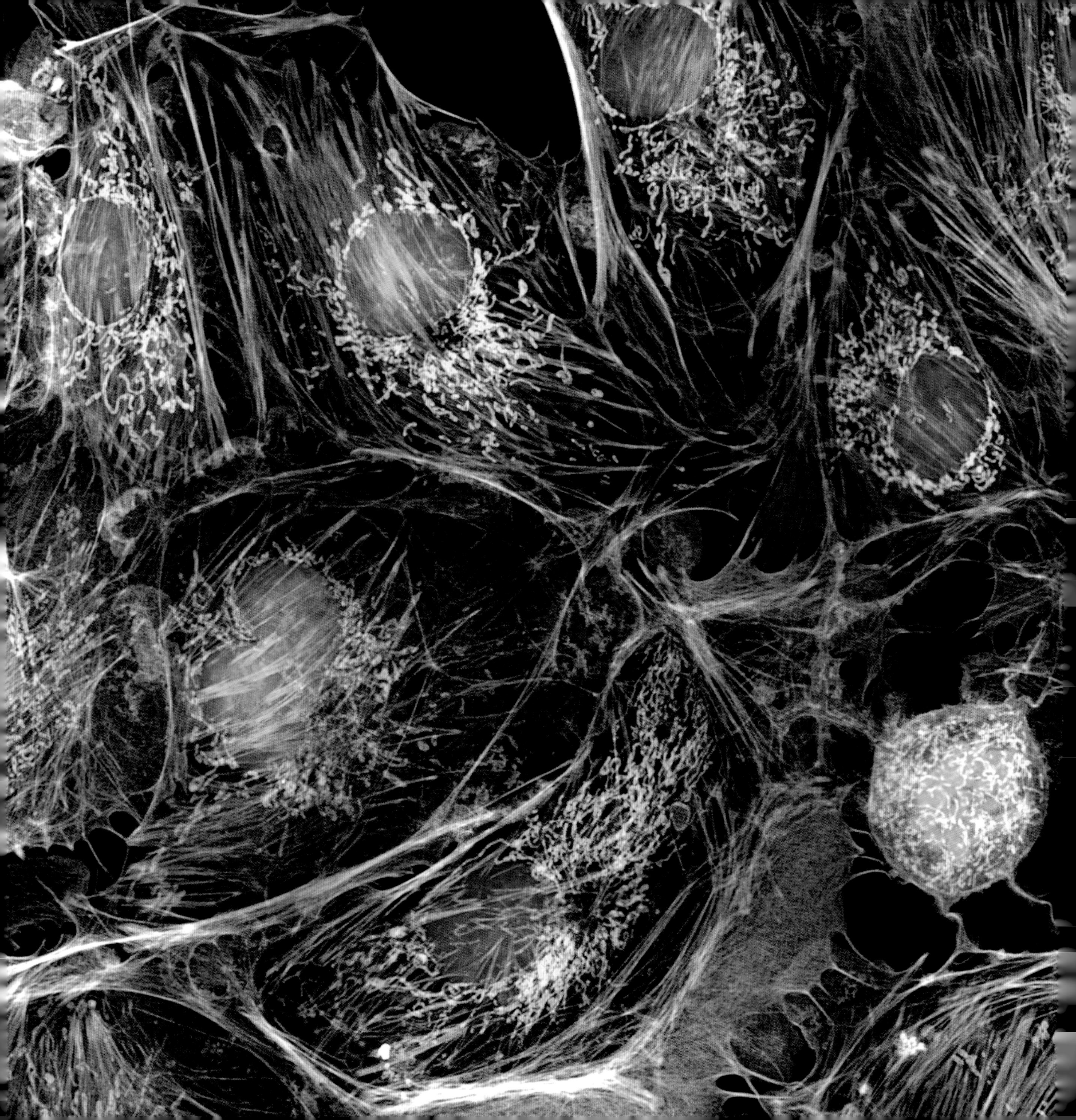

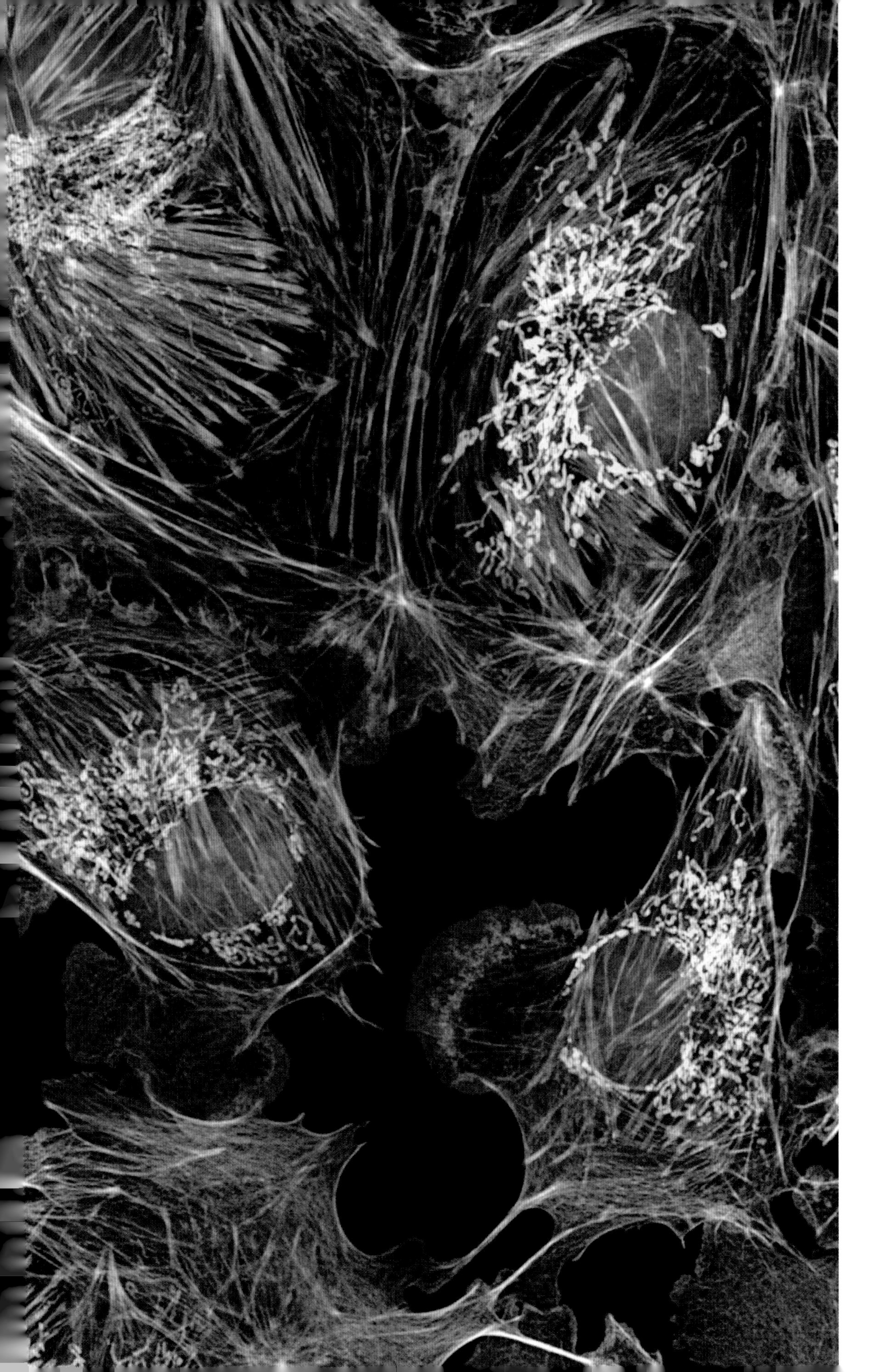

Previous page: Serotonin crystals (polarized light micrograph)
Serotonin is released by blood platelets during clot formation, where it causes the constriction of blood vessels. It is an important neurotransmitter (a messenger of the nervous system) in the brain, and a lack of it has been shown to cause depression. It is this function which has led to the development of SSRI (selective serotonin reuptake inhibitor) antidepressants, such as Prozac (fluoxetine). The hormone melatonin, which regulates our sleep cycle, is made from serotonin in the brain's pineal gland. (Magnification: unknown)

Left: Lung cells (immunofluorescence light micrograph)
A network of arteries and veins called the systematic circulation supplies most of the body with blood, but there are separate systems for the heart (coronary circulation) and the lungs (pulmonary circulation). These cells form the inner lining (the endothelium) of the pulmonary circulation. In this image, nuclei, which contain the cell's genetic information, appear blue; actin filaments, which are major components of the cytoskeleton, are white; mitochondria, which generate energy for the cell, are yellow. (Magnification: unknown)

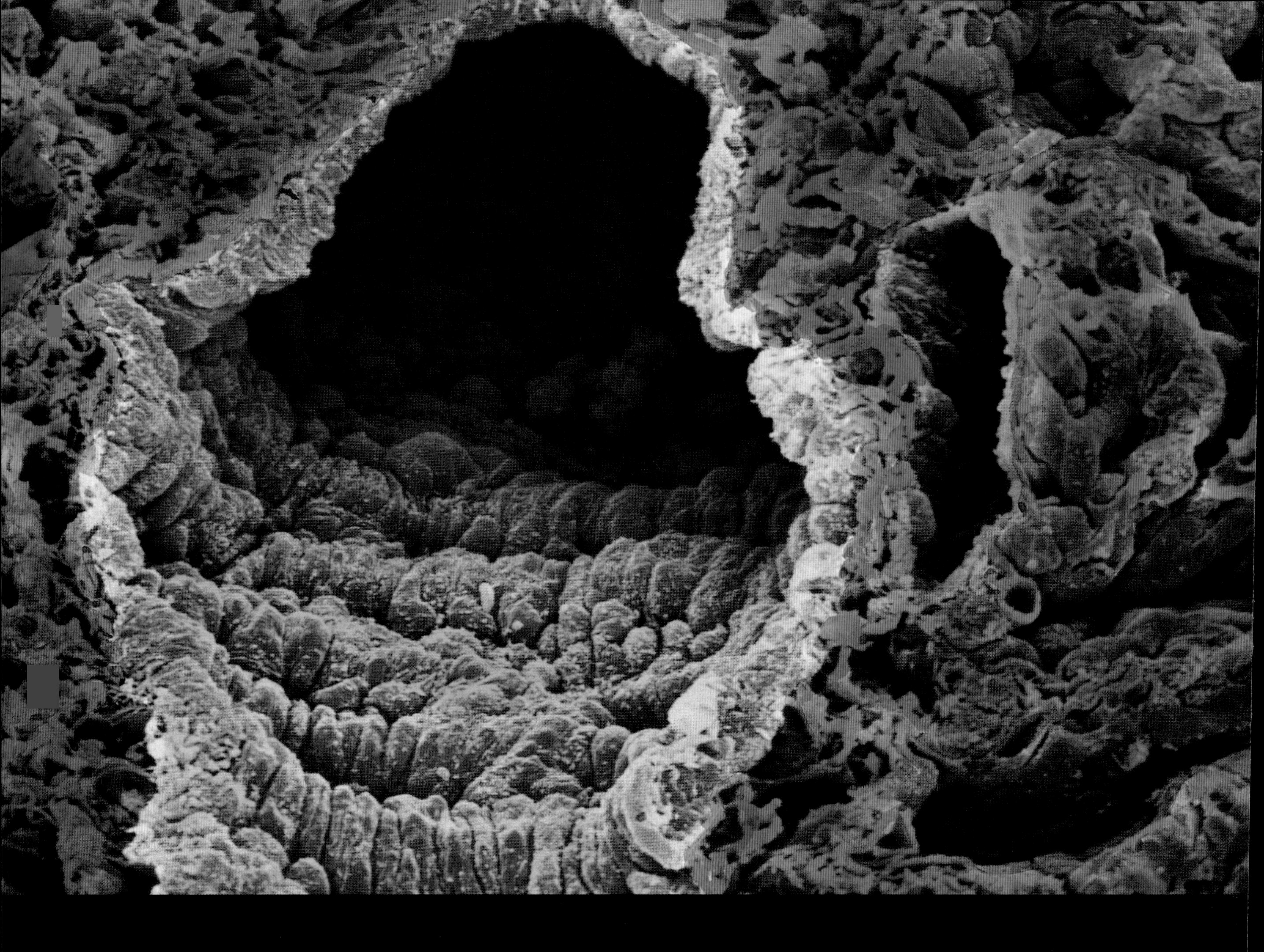

Above: Lung alveoli (coloured scanning electron micrograph)
At the centre of this image is an artery (yellow) supplying blood to the lungs, with the elongated cells (yellow-green) of the endothelium, the artery's inner lining. Immediately to its right, the elongated blue cavity is an arteriole, which distributes blood from the artery to even smaller blood vessels called capillaries (orange). The capillaries surround alveoli (the other blue cavities in this image), chambers in which carbon dioxide and oxygen are exchanged via the capillaries when we breathe. (Magnification: x900 at 6x7cm/2¼x2¾in size)

Right: Lung alveoli and bronchus (coloured scanning electron micrograph
When we inhale, air enters the lungs from the windpipe through bronchi (blue), which in turn supply tiny air sacs called alveoli, surrounded by capillary blood vessels (spongy yellow). Capillaries absorb oxygen from the alveoli and return carbon dioxide to them for us to exhale. Blood is supplied to capillaries from larger arteries (pink) via branches called arterioles (top centre-right). From the capillaries, re-oxygenated blood is then returned to the heart to be circulated around the body. (Magnification: x47 at 6x7cm/2¼x2¾in size)

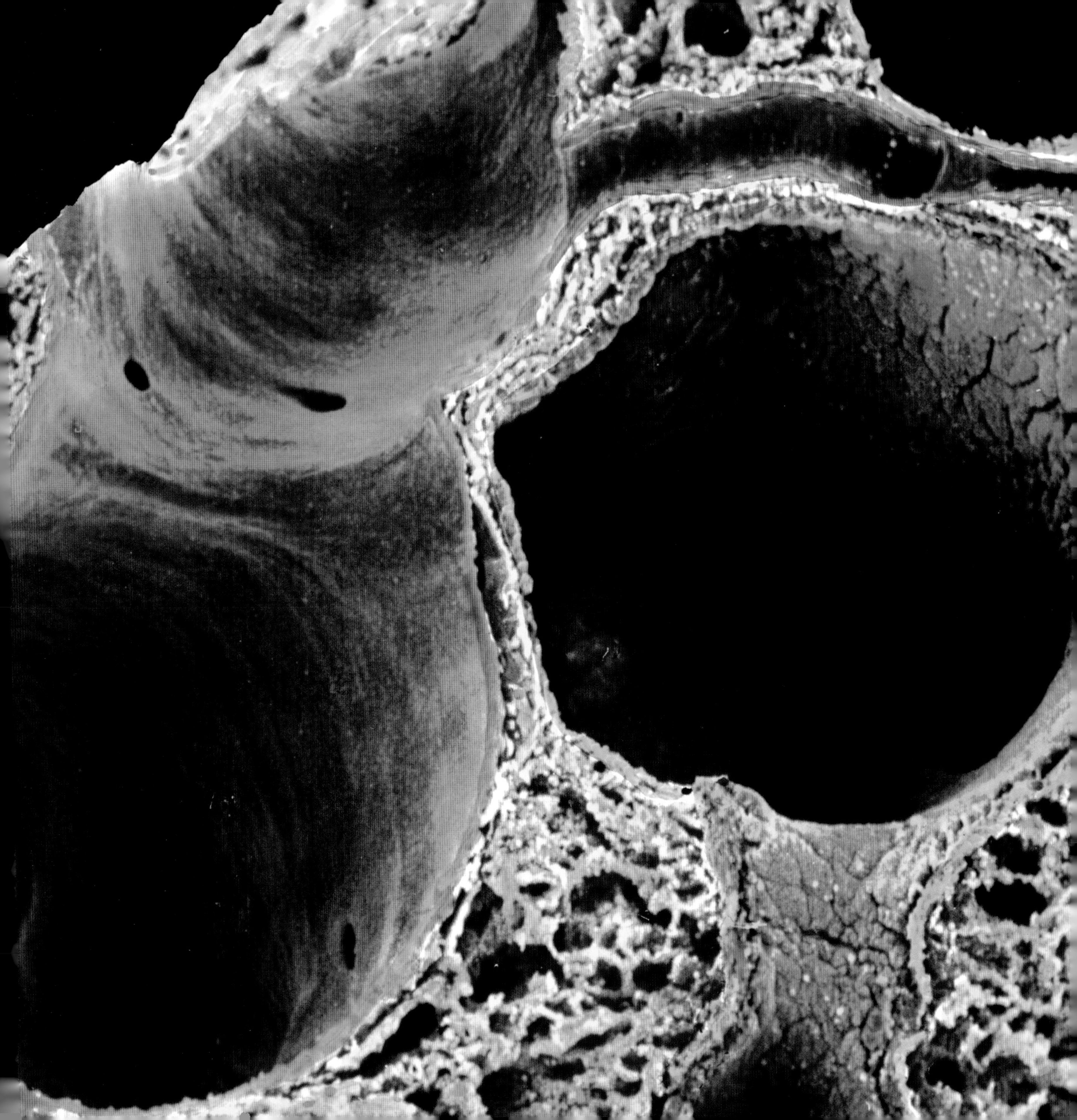

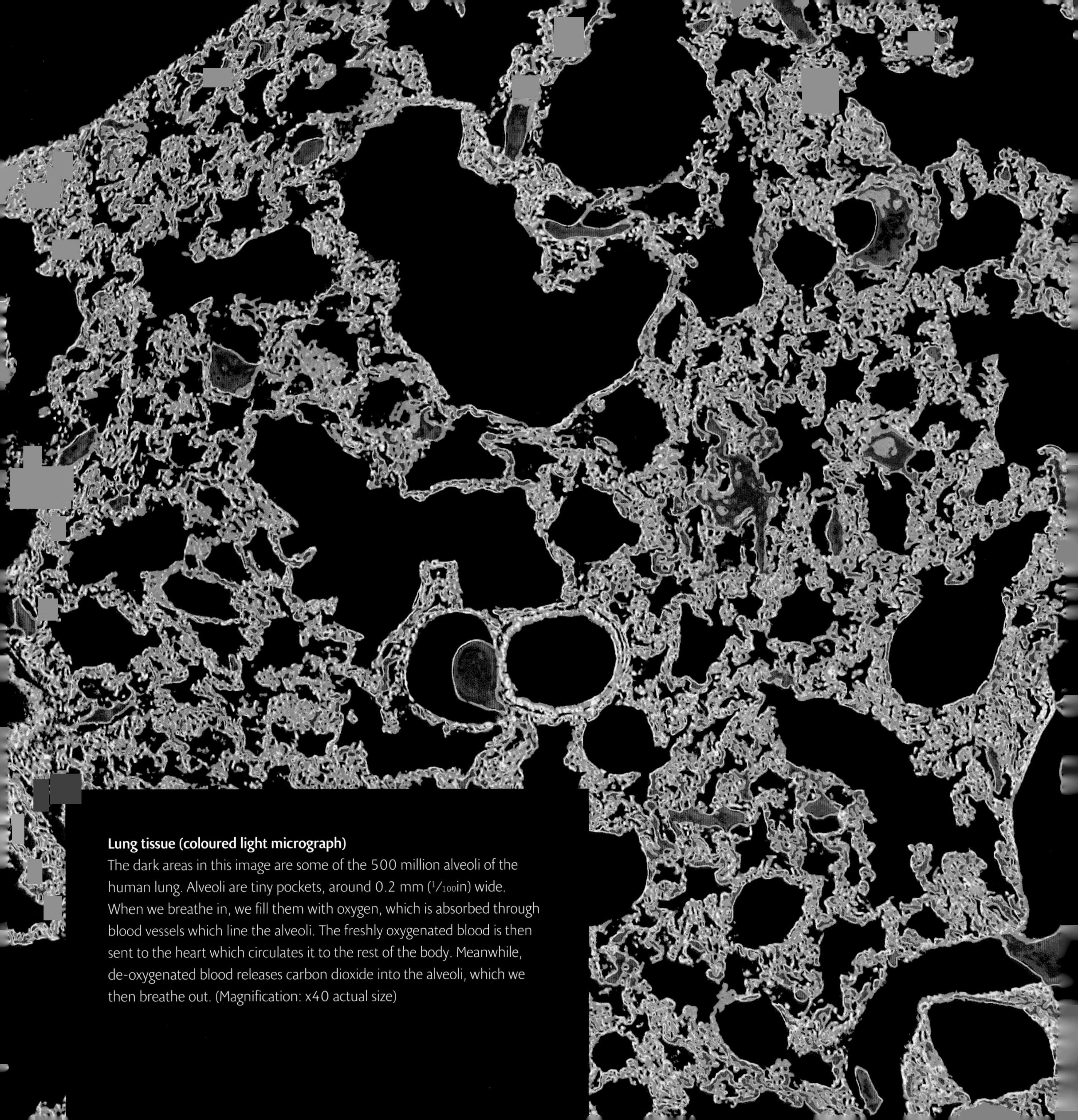

Lung tissue (coloured light micrograph)

The dark areas in this image are some of the 500 million alveoli of the human lung. Alveoli are tiny pockets, around 0.2 mm (1/100in) wide. When we breathe in, we fill them with oxygen, which is absorbed through blood vessels which line the alveoli. The freshly oxygenated blood is then sent to the heart which circulates it to the rest of the body. Meanwhile, de-oxygenated blood releases carbon dioxide into the alveoli, which we then breathe out. (Magnification: x40 actual size)

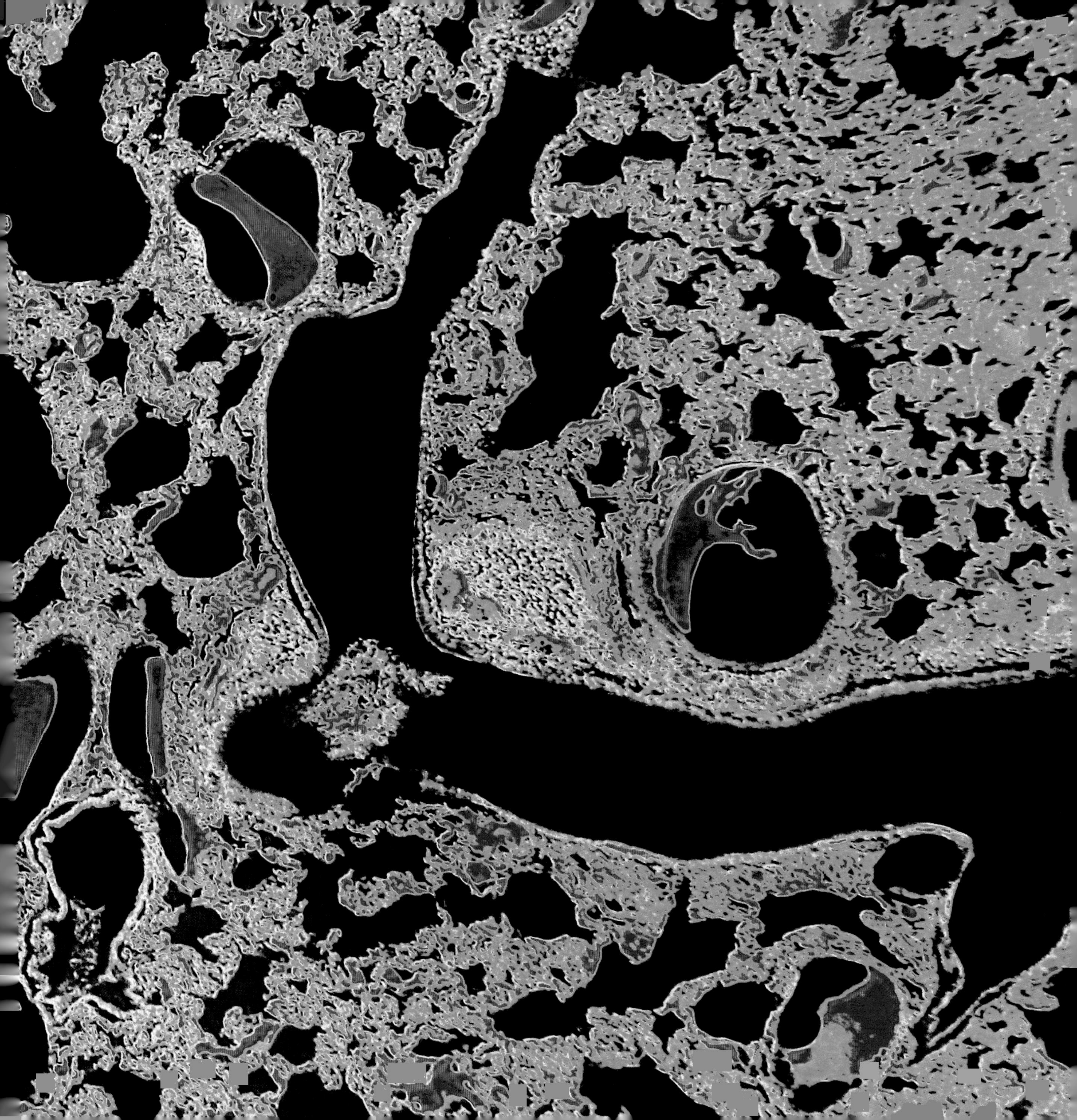

Left and above: Adrenaline crystals (polarized light micrographs)
Adrenaline, also called epinephrine, is normally present in blood in small quantities. It is a hormone produced in the adrenal glands above the kidneys. The glands are controlled by the hypothalamus, the part of the brain responsible for instinct and emotion. In times of stress more adrenaline is secreted into the bloodstream. It widens the airways of the lungs and constricts small blood vessels. This makes the muscles work harder and produces a 'fight or flight' response. Adrenaline used as a drug expands the bronchioles in acute asthma attacks, and stimulates the heart in cases of anaphylactic shock. (Magnification: x30 (left) and x50 (above) at 10x12.5cm/4x5in size)

Serotonin crystals (polarized light micrograph)

The human gut holds about 90% of the body's serotonin, which controls the digestive movement of the intestine. Elsewhere serotonin molecules act as messengers in the nervous system, where it has a role in memory and learning. It also regulates our mood, sleep and appetite; low levels of it may lead to depression. In the bloodstream it is absorbed and stored by platelets, which release serotonin to help with clotting. (Magnification: x4 actual size)

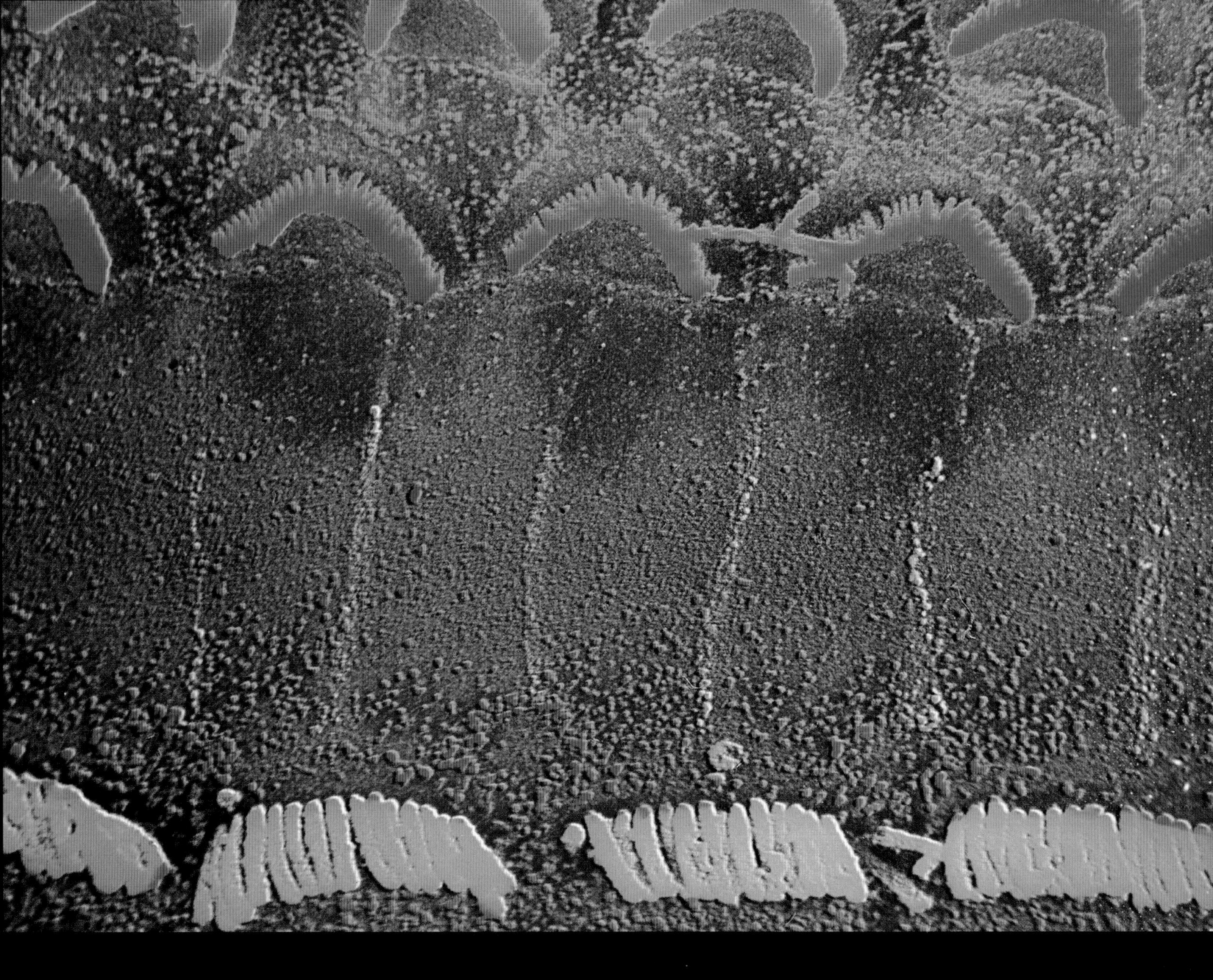

Above: Organ of Corti in the inner ear (coloured scanning electron micrograph)
The channels, membranes and bones of the outer and middle ear convey soundwaves to the inner ear. There, it is the organ of Corti (Corti was a nineteenth-century Italian anatomist) which hears the sounds. Hair cells line the organ, each supporting up to a hundred hair-like stereocilia arranged either in a line (yellow) or an inverted V (pink). Stereocilia convert the movement caused by soundwaves into electrical impulses that are transmitted to the brain. (Magnification: x1460 at 6x7cm/2¼x2¾in size)

Right: Balancing stone from inner ear (coloured scanning electron micrograph)
Our sense of balance is derived from a tiny stone in each ear, called an otolith (from Greek, literally 'ear-stone'). The stones are built up in the inner ear from deposits of calcium carbonate crystals, seen here on the surface of an otolith. Otoliths are attached to sensory hairs which are sensitive to gravity and acceleration. When the head tilts, the movement of the hairs sends nerve impulses to the brain, so that we can adjust our balance as required. (Magnification: unknown)

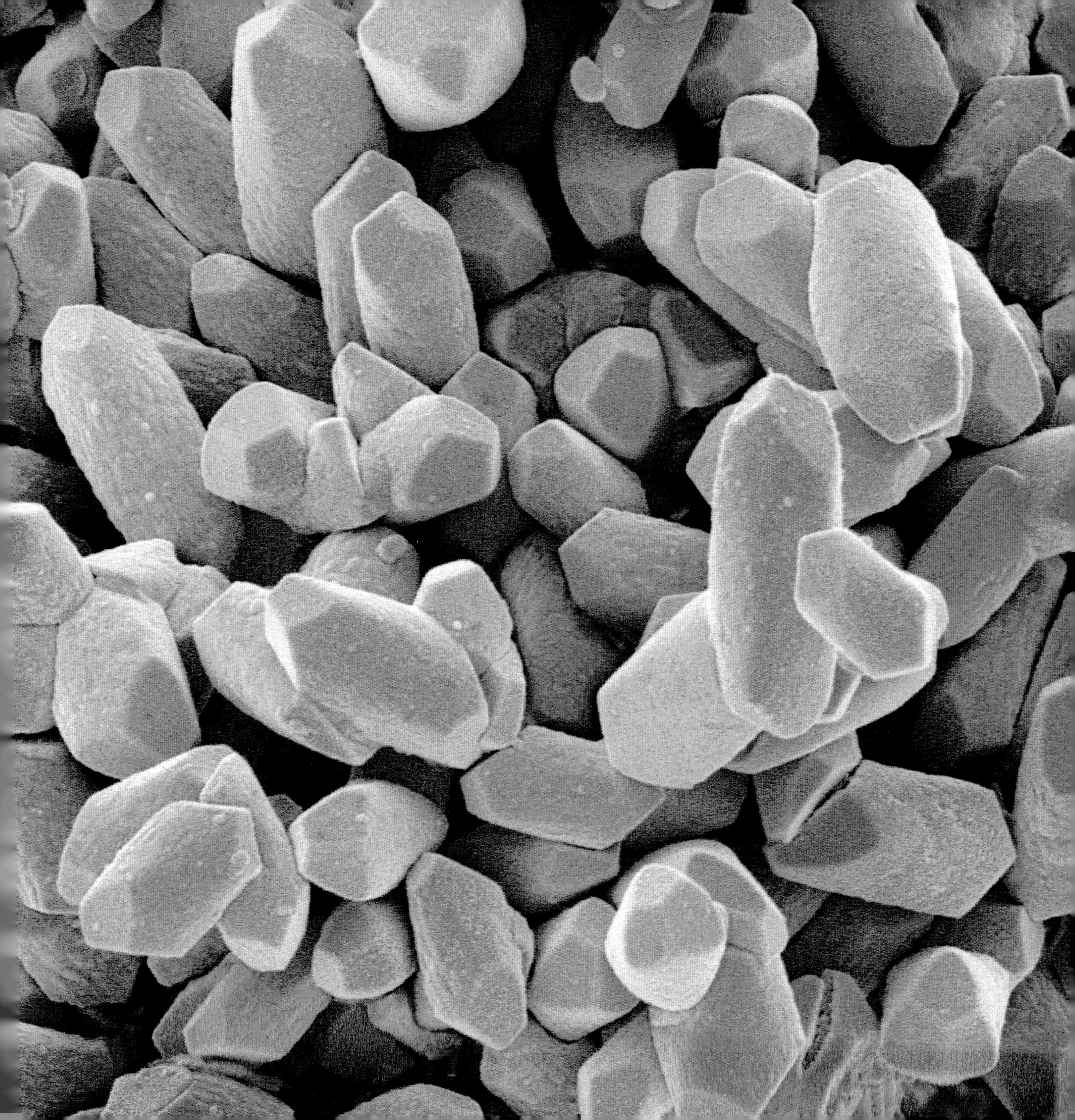

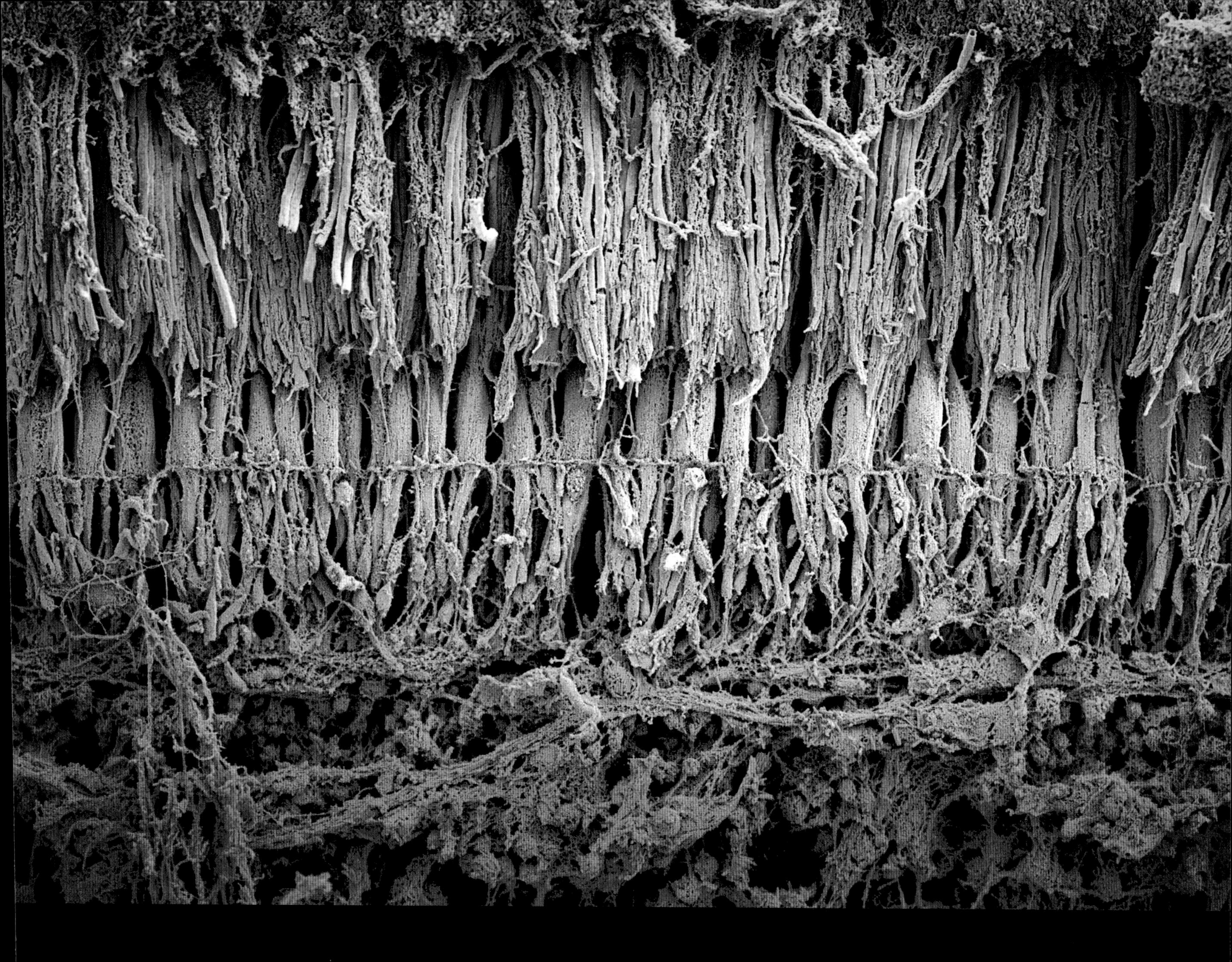

Left: Eye lens (coloured scanning electron micrograph)
The lens of the eye sits inside the eyeball immediately behind the iris (the coloured part). It focusses light onto the sensory retina which lines the eyeball. The lens is constructed of elongated cells arranged in concentric rings, seen here in flattened cross-section. It is held in place by minute ligaments attached to muscles around the iris. To focus on objects at different distances, the muscles pull on the ligaments and alter the curvature of the lens. (Magnification: x100 at 10cm/4in size)

Above: Retina layers (coloured scanning electron micrograph)
The retina of the human eye lines the inside of the eyeball and detects light focussed onto it by the lens. Approaching the retina from below in this image, light passes through a translucent layer of blood vessels and nerves on the retina (the lower third of the image) before striking the tightly packed photoreceptors (light-receiving cells) attached to the epithelium (the back wall of the retina, at the top of the picture). (Magnification: x480 at 10cm/4in size)

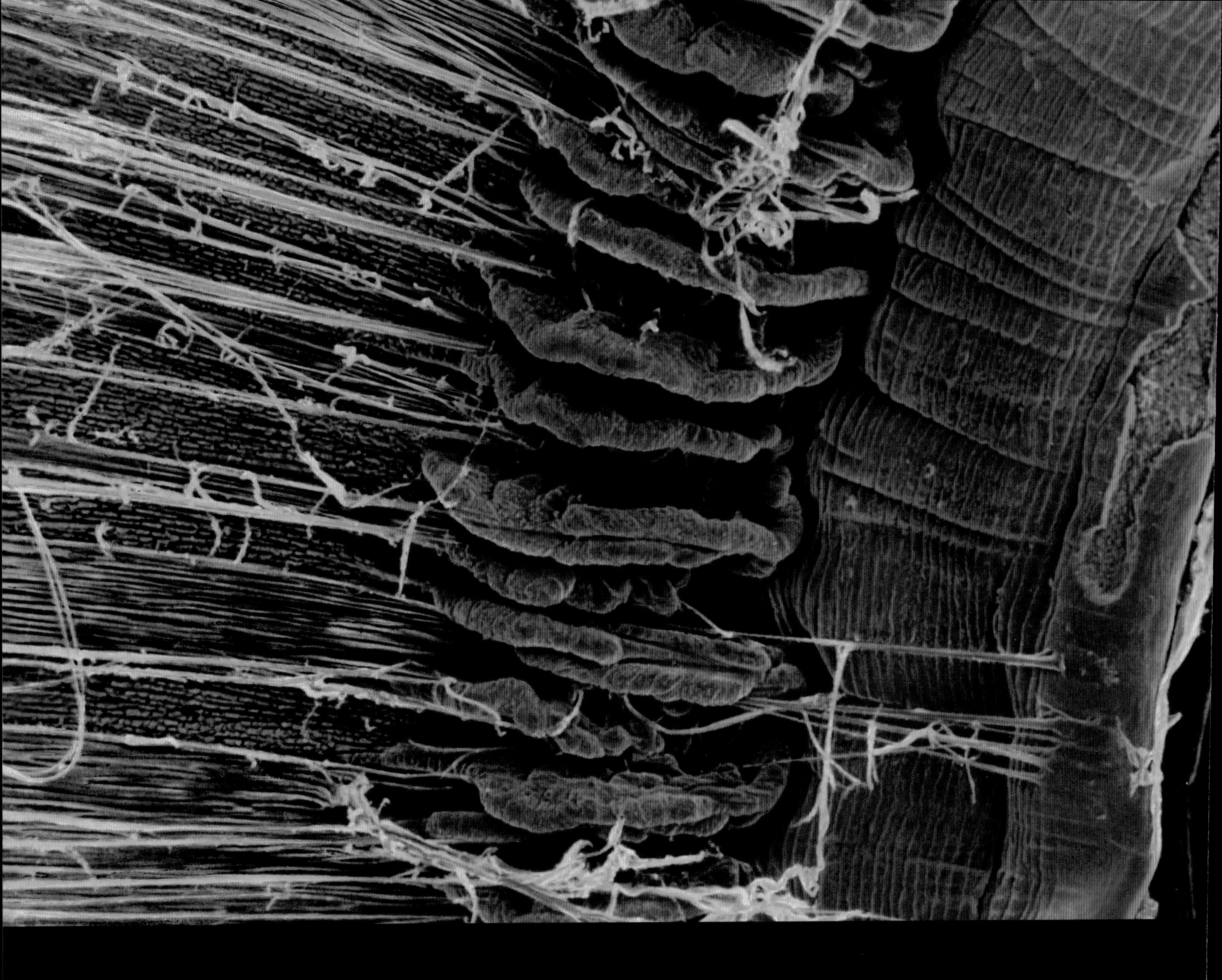

Above: Iris of the eye (coloured scanning electron micrograph)
The eye has a number of mechanisms to control how we see. At far lower right is the edge of the pupil (blue), the hole that allows light into the eye. Beside it, the iris (pink) controls the size of the pupil and therefore how much light will enter. The band of folds down the centre are the ciliary processes (red), muscles which pull on thin zonule filaments (yellow and green) to alter the shape and focus of the lens. (Magnification: unknown)

Right: Retina (coloured scanning electron micrograph)
This is a resin cast of the retina, the light-sensitive membrane that lines the back of the eye. Blood vessels (pink, orange) radiate out from the optic disc (yellow), the area where the optic nerve and blood vessels enter the eyeball. This part of the retina has no light-sensitive cells and is the cause of our so-called 'blind spot'. The blood vessels have been injected with resin and revealed by removing the surrounding tissue. (Magnification: unknown)

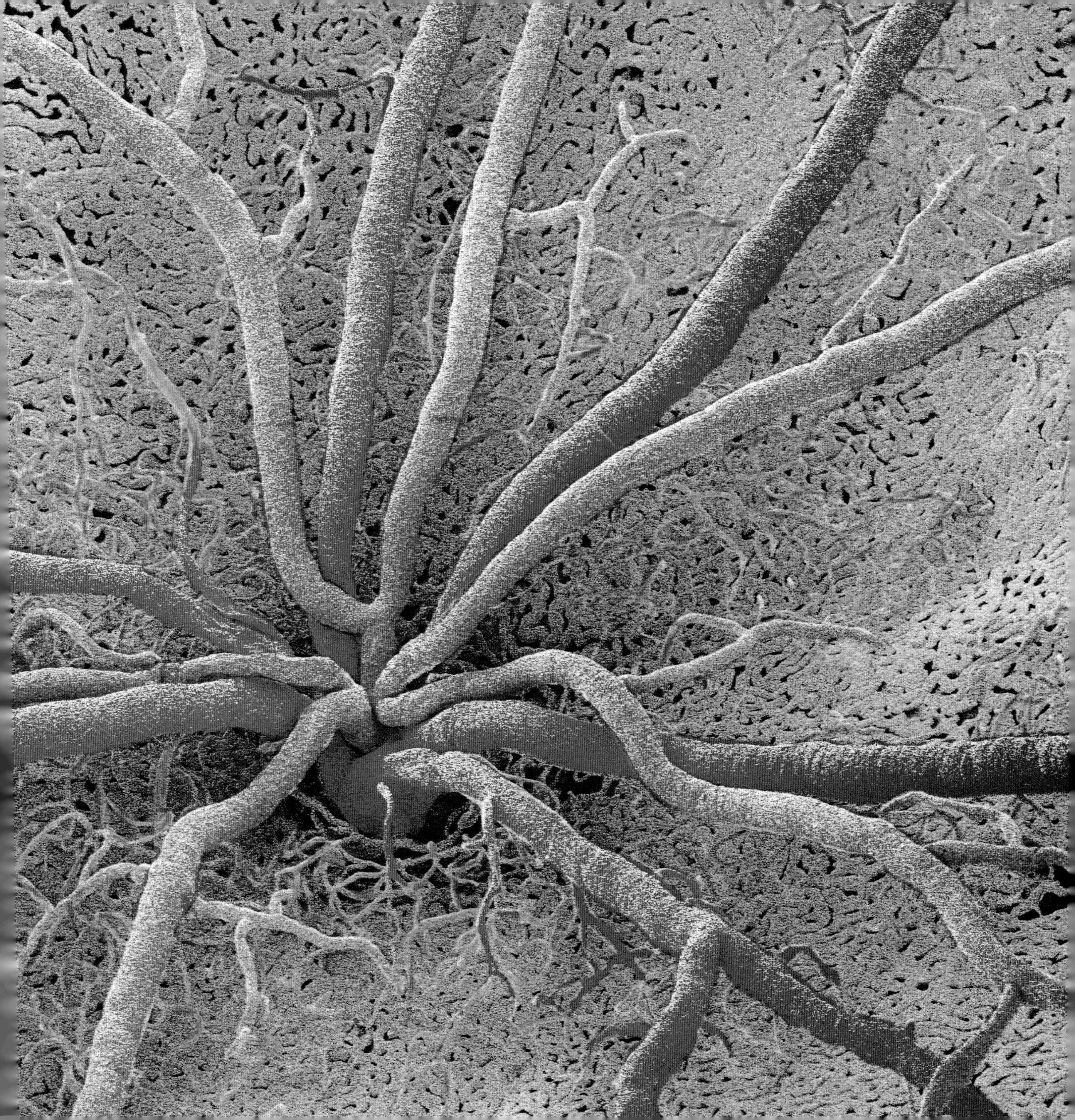

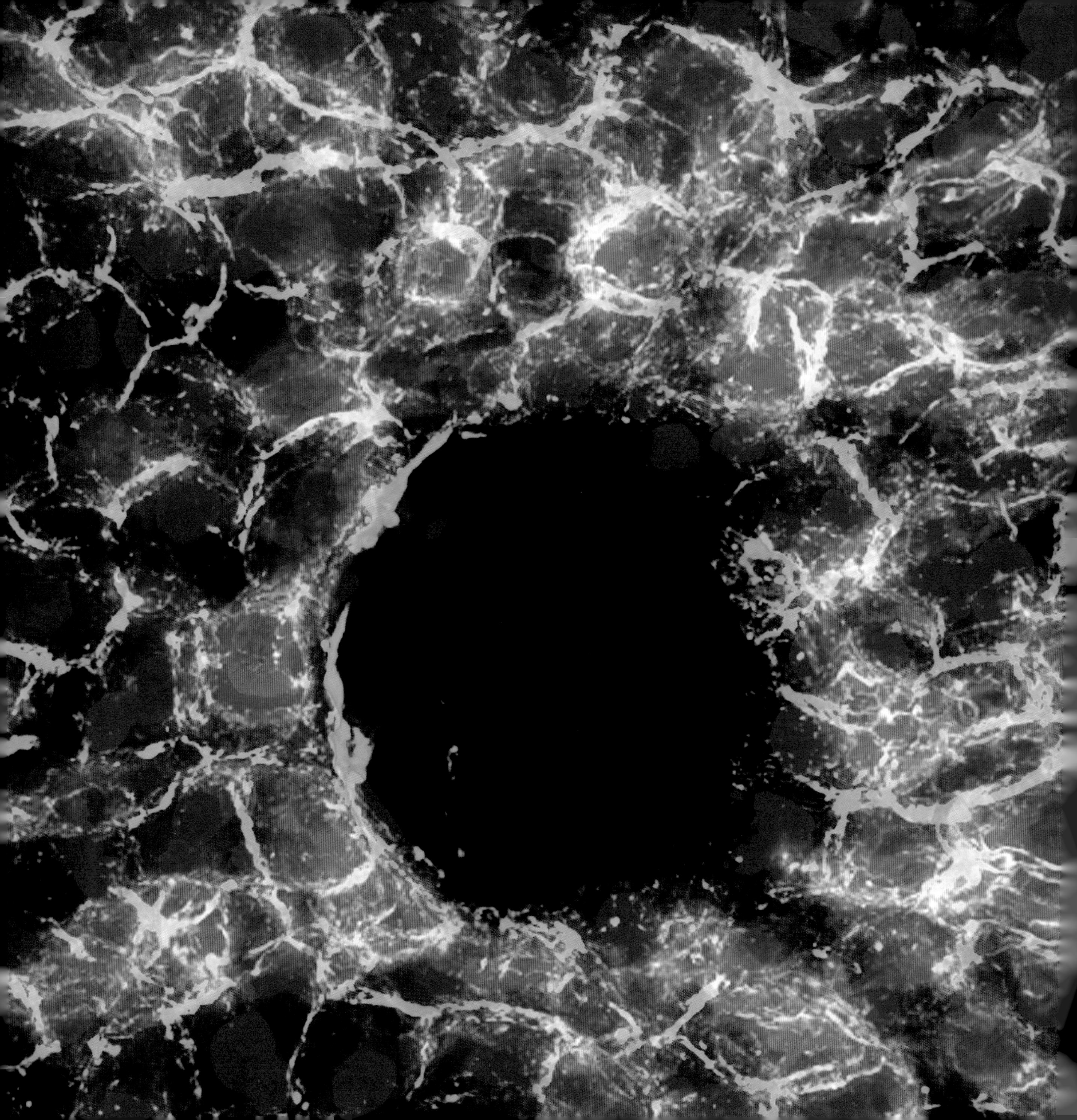

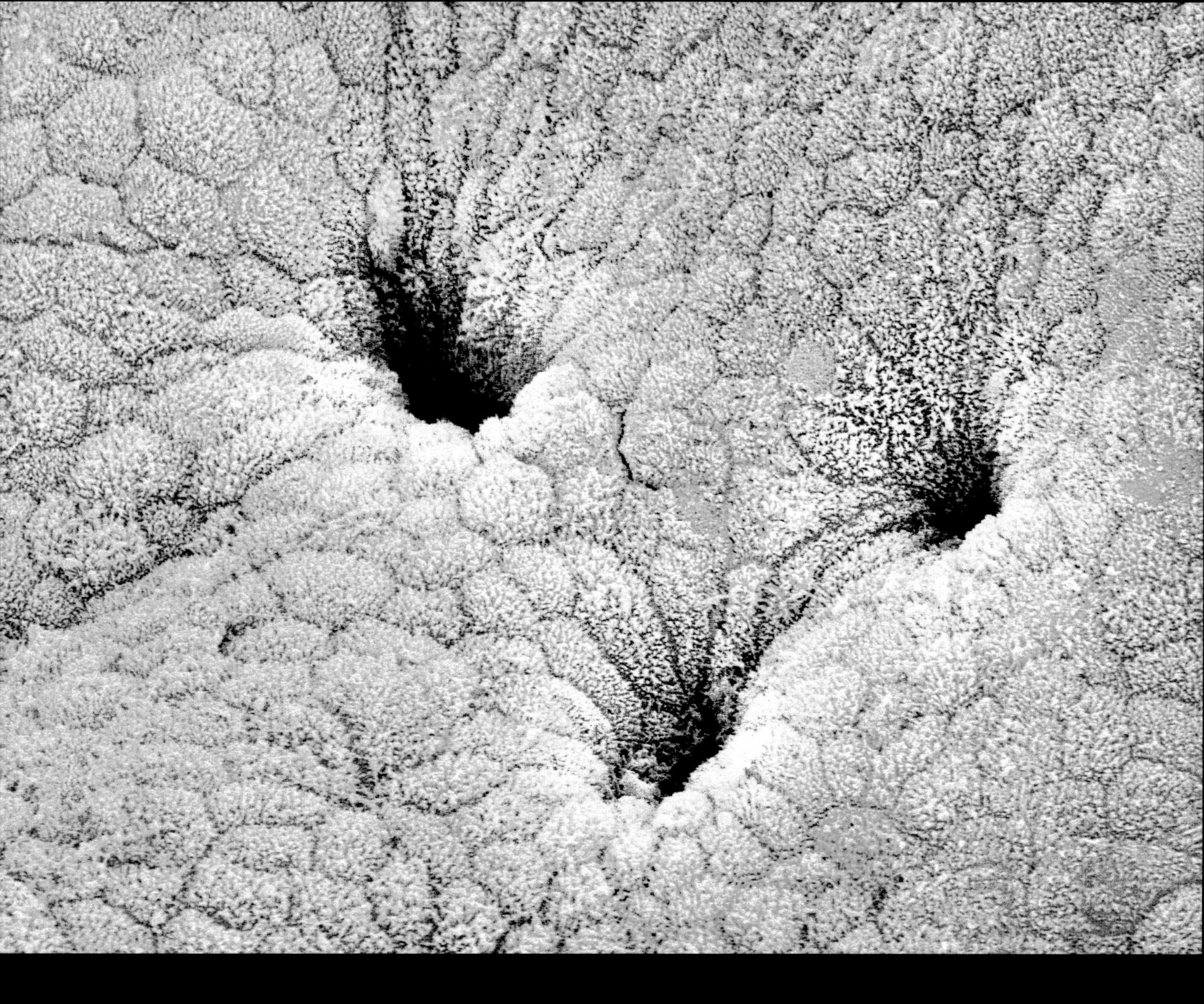

Left: Liver tissue (fluorescence deconvolution light micrograph)
The liver is the body's largest organ, storing nutrients and cleaning the blood of toxins and other waste products. This liver tissue sample includes a cross-section through one of the organ's central veins (black). Each vein forms the core of a liver lobule, a cluster of processing cells (hepatocytes) which receive the blood that has circulated through the liver, returning it to the body. The cells' structural proteins are tagged with red and green fluorescent markers; the nuclei appear blue. (Magnification: unknown)

Above: Gallbladder surface (coloured scanning electron micrograph)
The gallbladder stores and concentrates bile produced by the liver and delivers it to the small intestine where it helps the body to digest fat. At the centre of this image are three bile-emitting glands (yellow). The mound-like cells of the surface (columnar epithelium) are covered in tiny hair-like projections (microvilli). This arrangement absorbs water to make the bile more concentrated and effective in the digestive system. (Magnification: unknown)

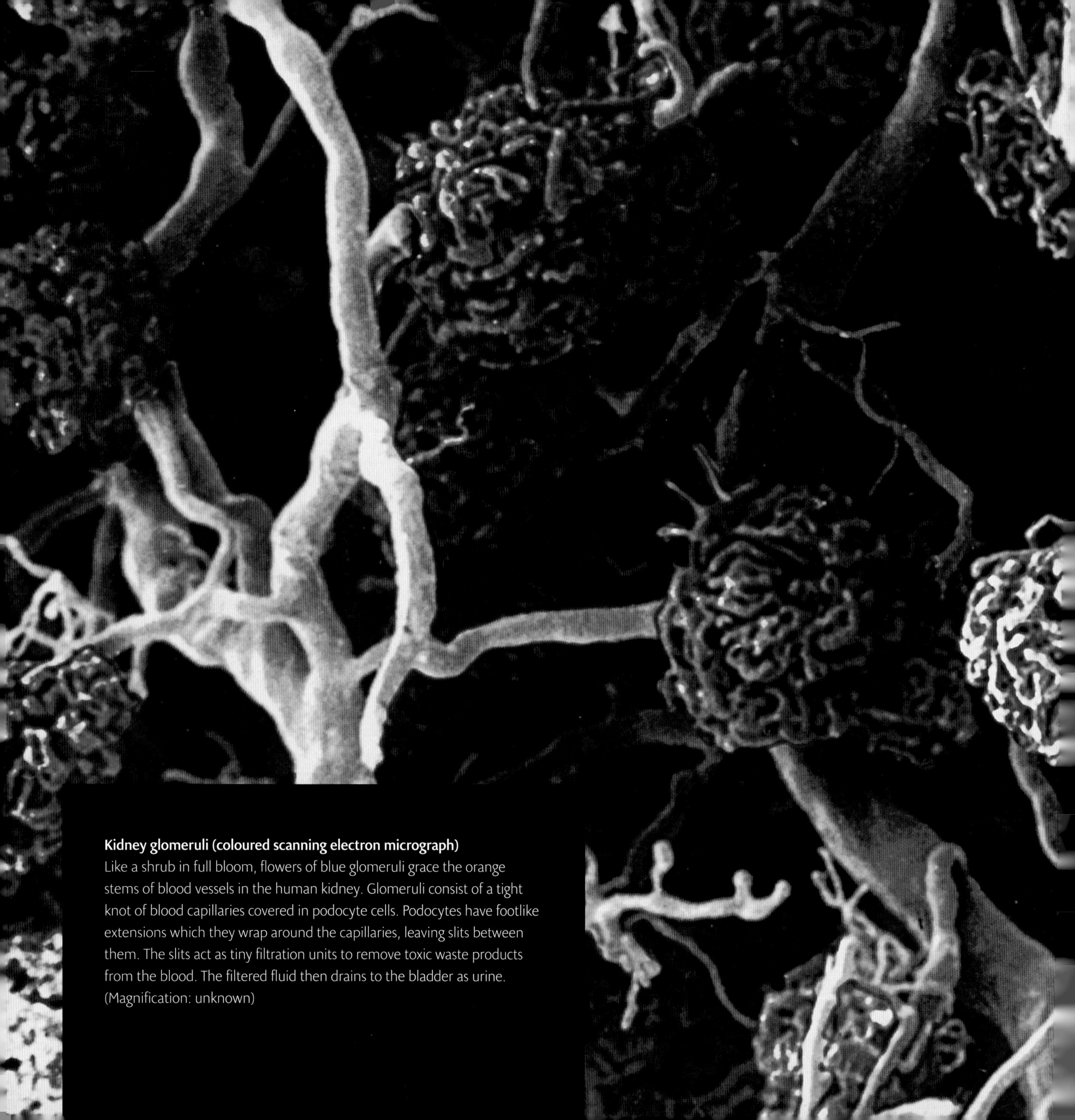

Kidney glomeruli (coloured scanning electron micrograph)
Like a shrub in full bloom, flowers of blue glomeruli grace the orange stems of blood vessels in the human kidney. Glomeruli consist of a tight knot of blood capillaries covered in podocyte cells. Podocytes have footlike extensions which they wrap around the capillaries, leaving slits between them. The slits act as tiny filtration units to remove toxic waste products from the blood. The filtered fluid then drains to the bladder as urine. (Magnification: unknown)

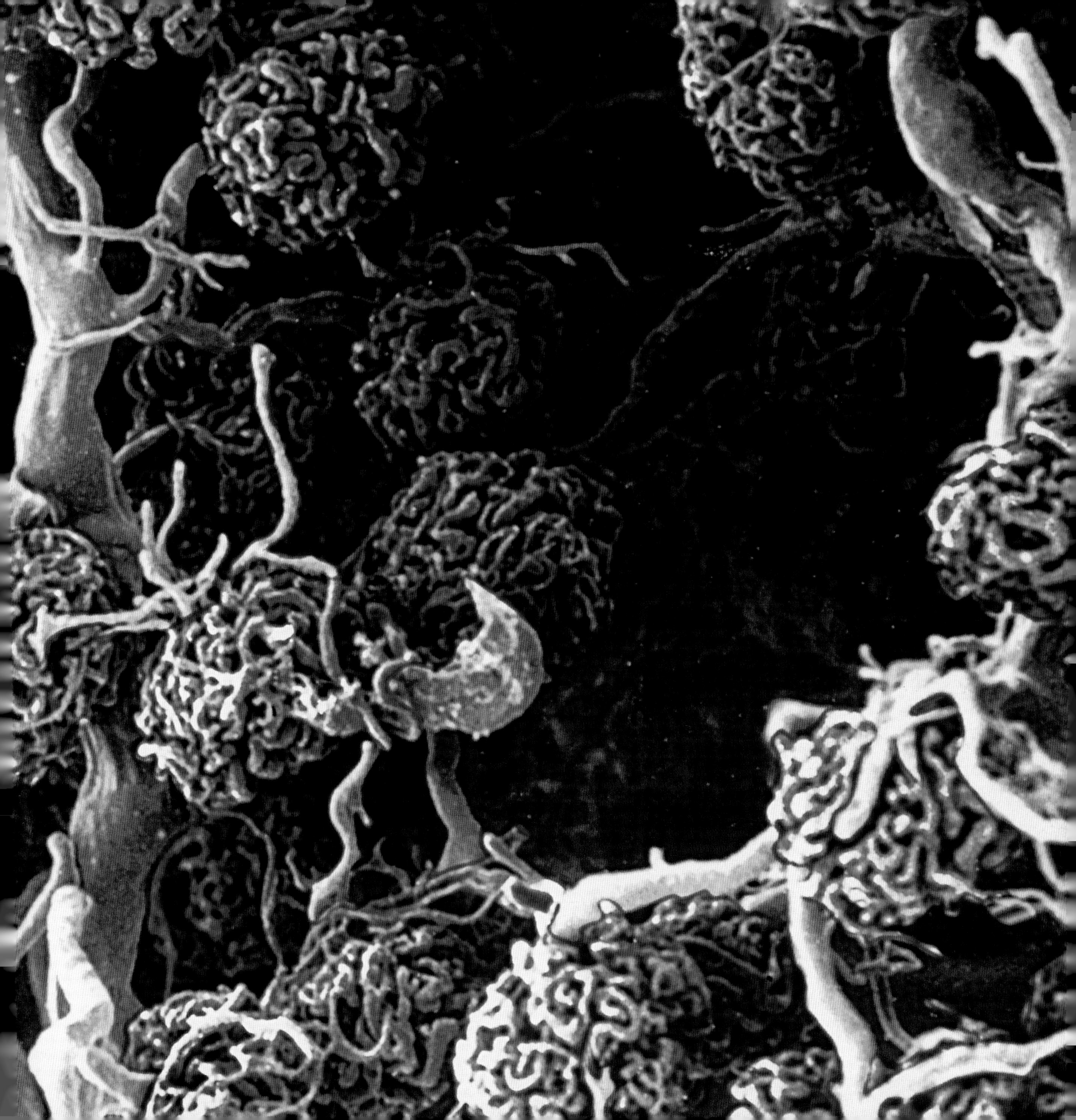

Kidney glomeruli (coloured scanning electron micrograph)

Each kidney has hundreds of glomeruli that filter toxic waste from the blood. The core of a glomerulus is a tight ball of capillaries, the smallest blood vessels. A mesh of podocyte cells clings to the capillaries with footlike extensions, filtering the blood. Waste fluid drains to the bladder as urine. In this resin cast, podocytes and the capsule which encloses the glomerulus have been removed to reveal the capillaries (pink) and the larger blood vessels (orange) which supply them. (Magnification: unknown)

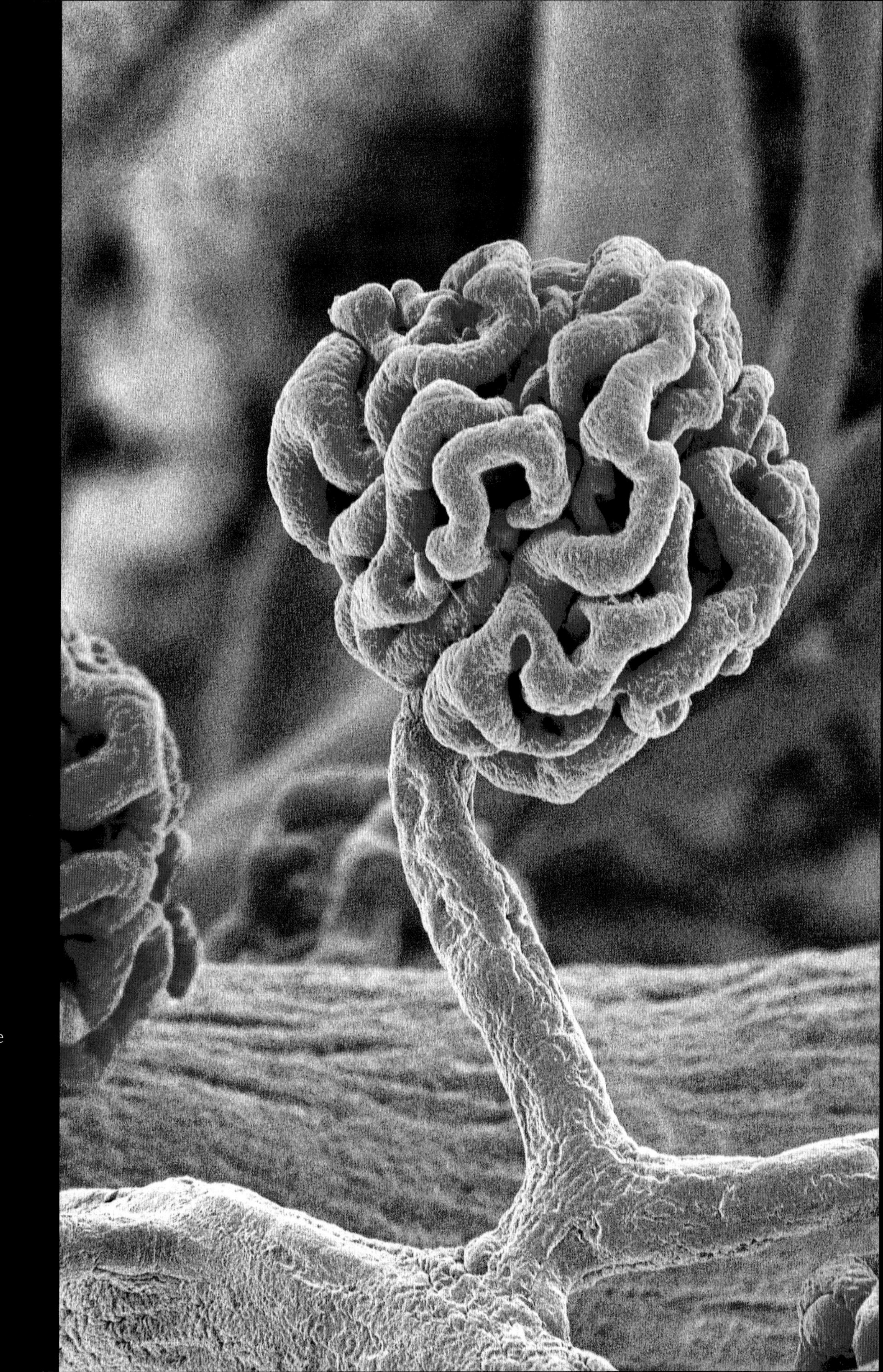

Kidney glomerulus (coloured scanning electron micrograph)

A glomerulus is a filtration unit in the kidney which removes toxic waste from the blood and disposes of it in urine. It is contained in a sac called Bowman's Capsule which holds the urine before it passes to the bladder. In this image the blood supply of the glomerulus has been injected with resin and the surrounding tissue, including Bowman's Capsule, removed to leave only the capillaries of the glomerulus and the larger blood vessel which supplies them. (Magnification: unknown)

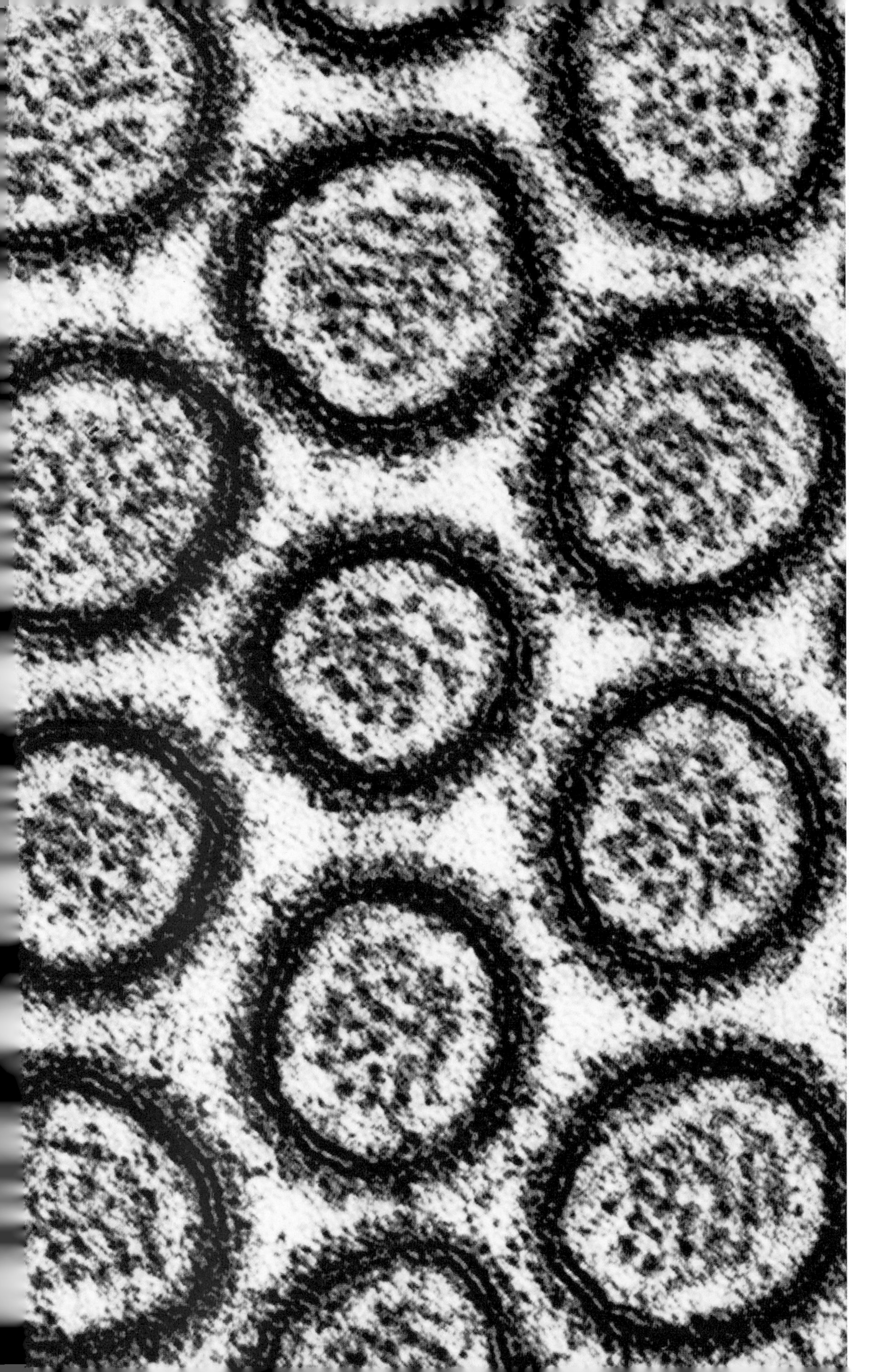

Intestinal microvilli (coloured transmission electron micrograph)

Villi are finger-like projections that line the wall of the small intestine. Microvilli, as the name implies, are even smaller hair-like projections extending from each villus (seen here in circular cross-section). They consist of a double membrane structure around a core of actin and other structural proteins. The microvilli increase the surface area of the intestine walls to maximize the absorption of nutrients from digested food. (Magnification: x22500 at 10cm/4in size)

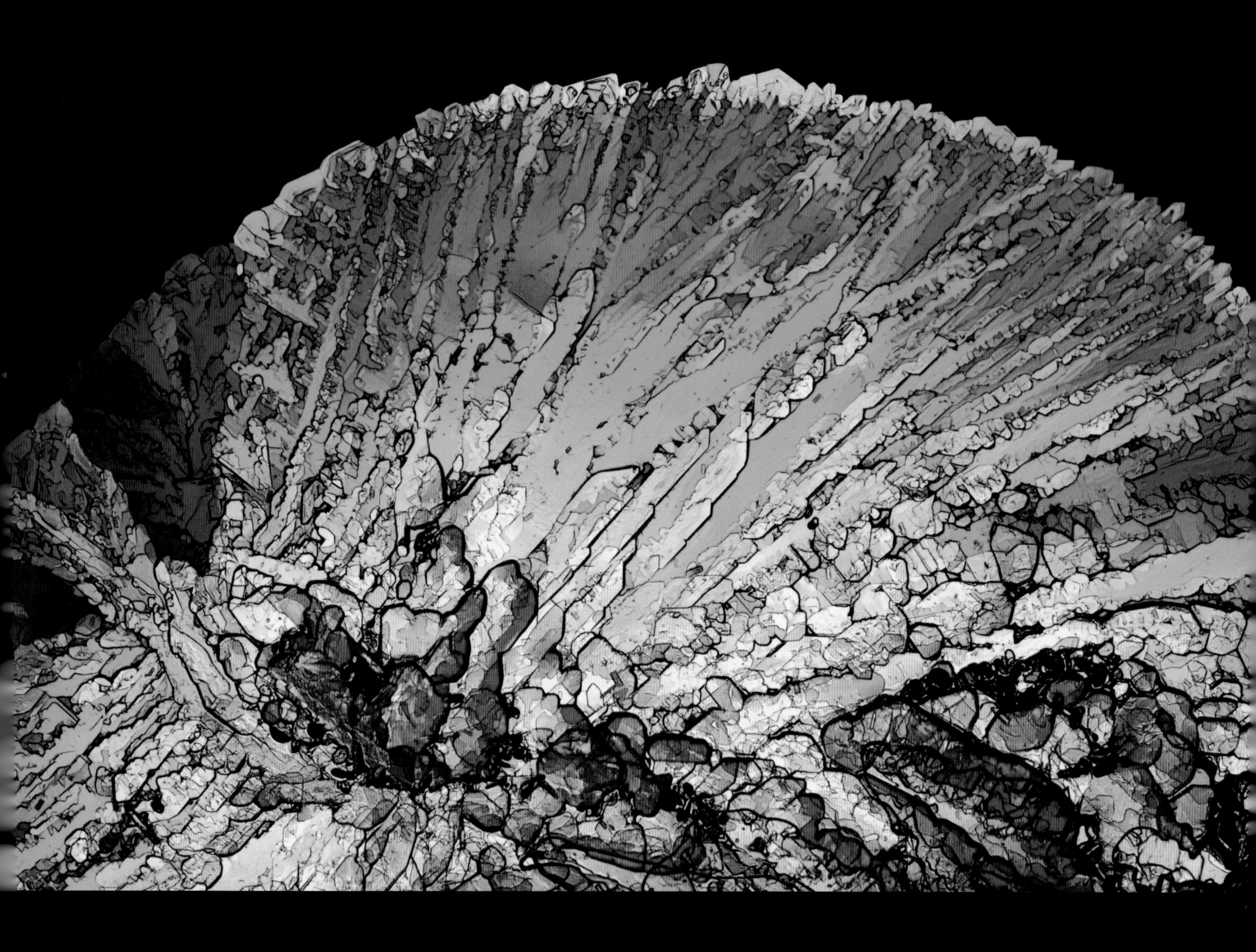

Left: Insulin crystals (polarized light micrograph)
These hexagonal crystals are of the hormone insulin. Insulin is produced in the pancreas, and its function is to regulate blood sugar levels. Insufficient production of insulin leads to an accumulation of glucose in the blood, and can cause Type 1 diabetes. Type 2 diabetes can occur when there is plenty of insulin, but the body's cells do not respond properly to it. A third type, gestational diabetes, occurs in pregnant women who produce high levels of blood glucose. (Magnification: x117 at 6x7cm/2¼x2¾in size)

Above: Insulin crystals (polarized light micrograph)
The insulin molecule is made up of two chains of amino acids which are held together by two sulphur bonds. It is a hormone produced by cells in areas of the pancreas called the islets of Langerhans (named after Paul Langerhans, a nineteenth-century German anatomist). Its production is stimulated by rising blood sugar levels, which it regulates by encouraging liver cells to store glucose. (Magnification: unknown)

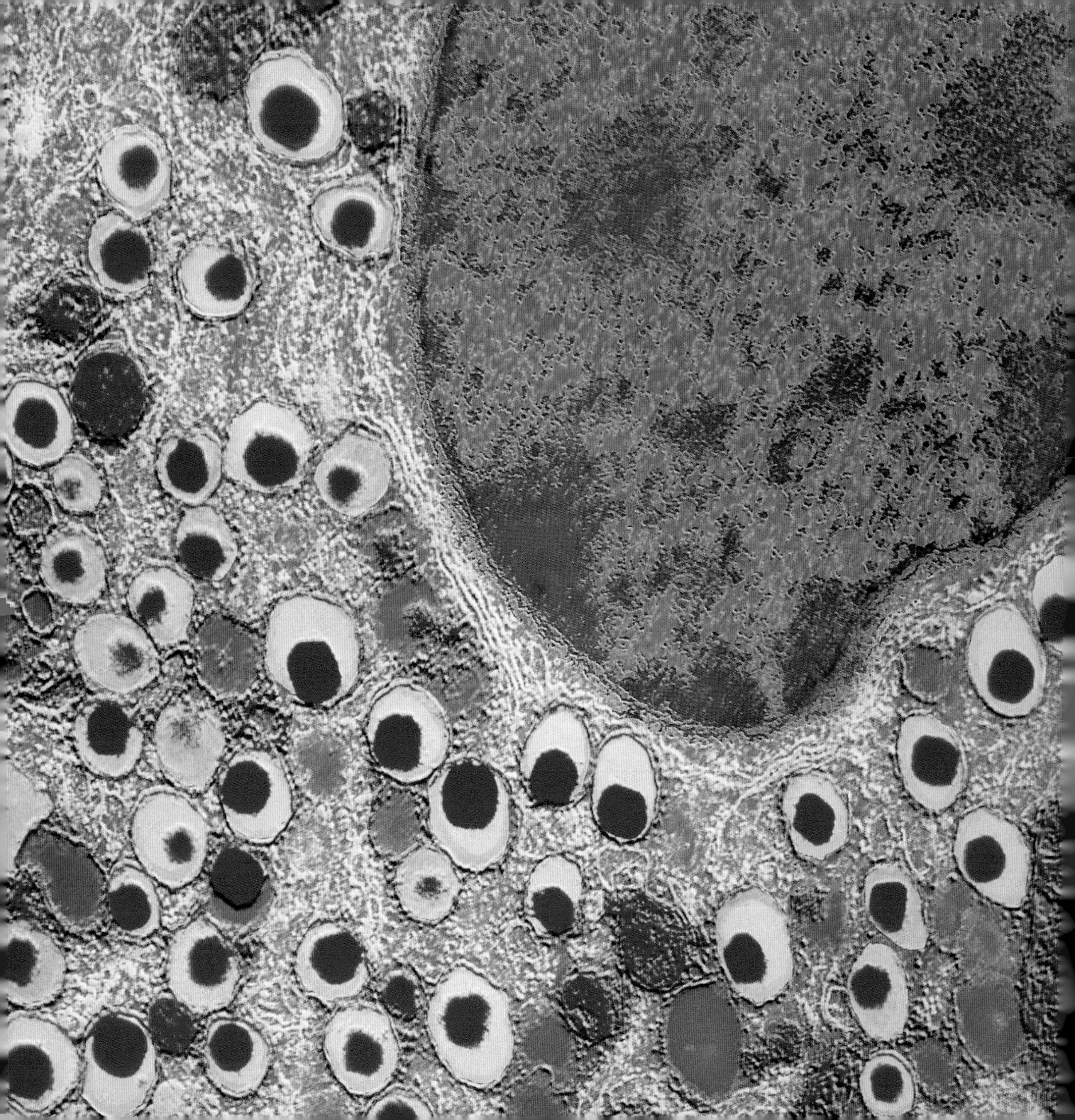

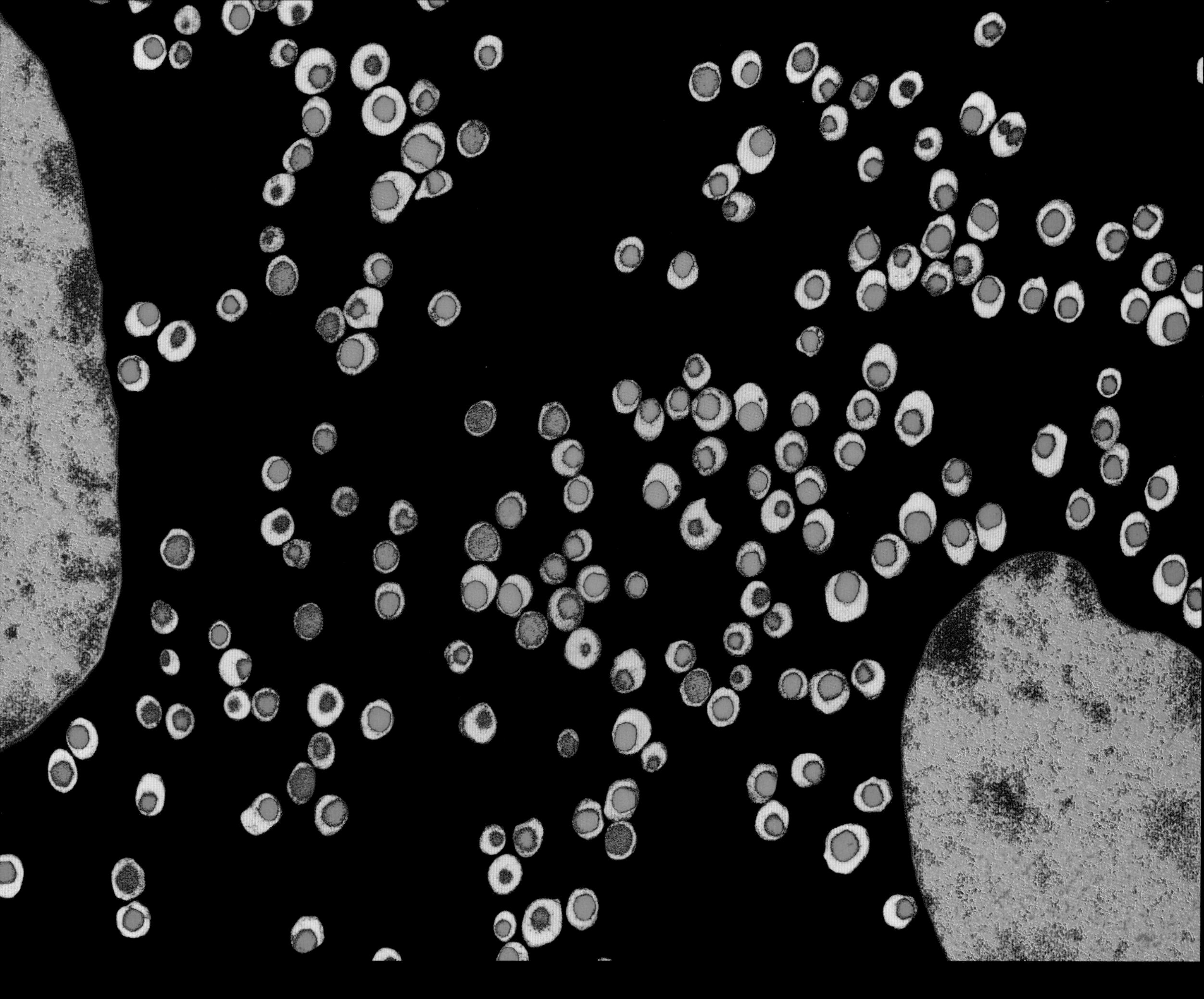

Left: Islet of Langerhans (colour transmission electron micrograph)
This image is a cross-section through a cell in one of a million islets of Langerhans in the pancreas. The islets of Langerhans secrete hormones into the blood – mostly insulin and glucagon, which control blood sugar levels. At top right (mauve and green) is part of the nucleus of this cell. The red spots in white spaces are secretory granules, each enclosed in a red membrane. These contain a dense, granular core of hormone. (Magnification: x4800 at 6x4.5cm/2¼x1¾in size)

Above: Islet of Langerhans (coloured transmission electron micrograph)
This image shows parts of two cells in a hormone-secreting islet of Langerhans, one of approximately one million islets of Langerhans in the human pancreas. The large objects far left and lower right are parts of the nuclei of these cells. The cells contain circular secretory granules which contain a dense granular core of hormone (green) separated from the enclosing membrane by a clear space (yellow). The membrane separating the two cells is not shown here. (Magnification: x8700 at 10cm/4in size)

Fimbriae of a fallopian tube (coloured scanning electron micrograph)
In the female body there are two egg-producing organs called ovaries. The fallopian tube is the channel from each through which a newly released egg cell passes on its way to the uterus. Fimbriae are the folds at the opening of the fallopian tube next to each ovary, which help to guide a newly-released ovum into the tube itself. It is in the fallopian tube that any fertilization of the egg normally occurs. (Magnification: x10 at 10cm/4in size)

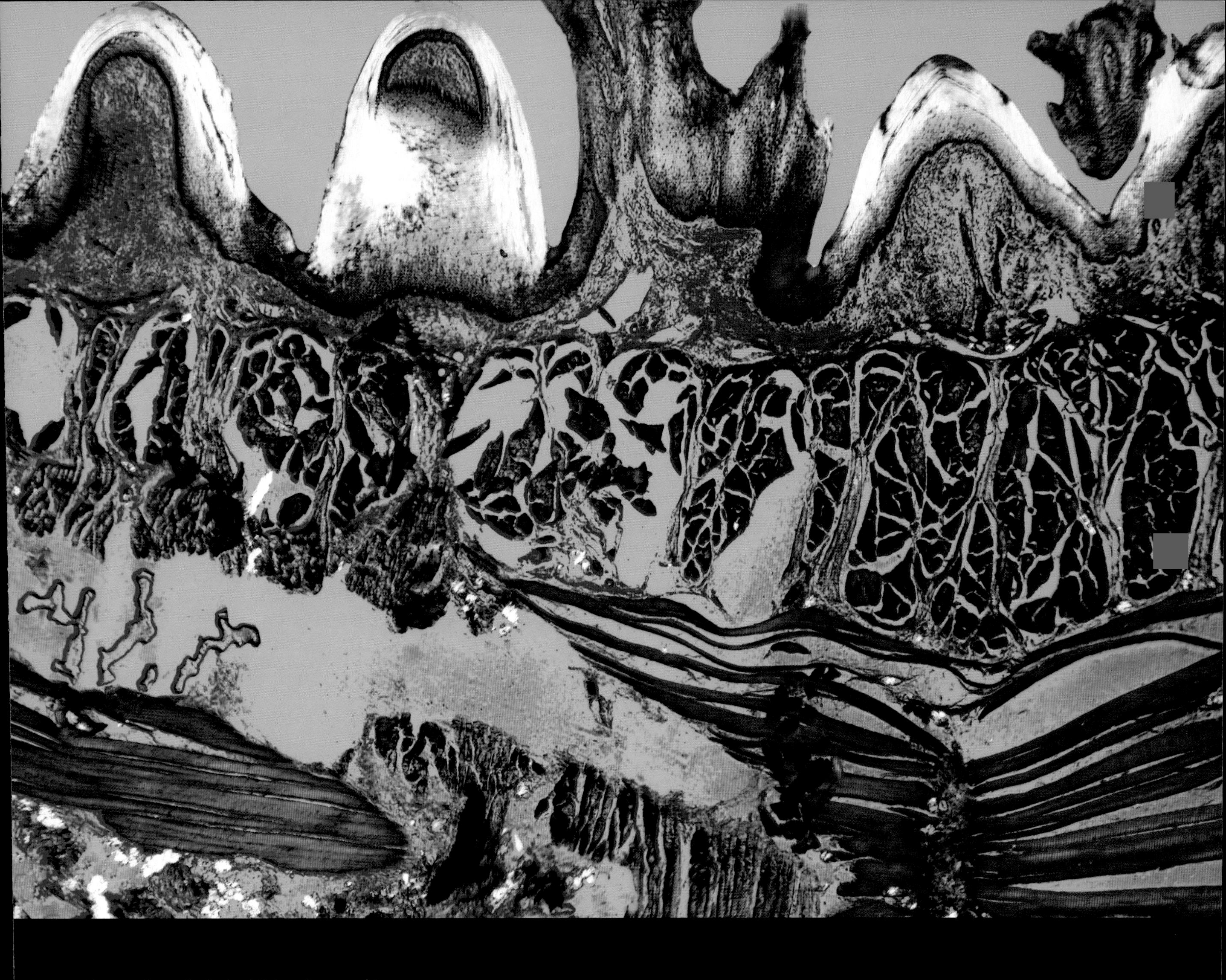

Tongue tissue (polarized light micrograph)
In this cross-section, the tongue's surface is at the top. On it are three rounded ducts (fungiform papillae), and one jagged (filiform) papilla. Below the papillae are layered blocks of muscle arranged vertically (light green), across the tongue (deep green) and along its length (red). The tongue has a covering of keratinized stratified skin (yellow), which protects it from dehydration and abrasion. Tongues also have circular (circumvallate) and leaf-shaped (foliate) papillae, not seen here; all four kinds carry taste buds. (Magnification: x100 at 10cm/4in size)

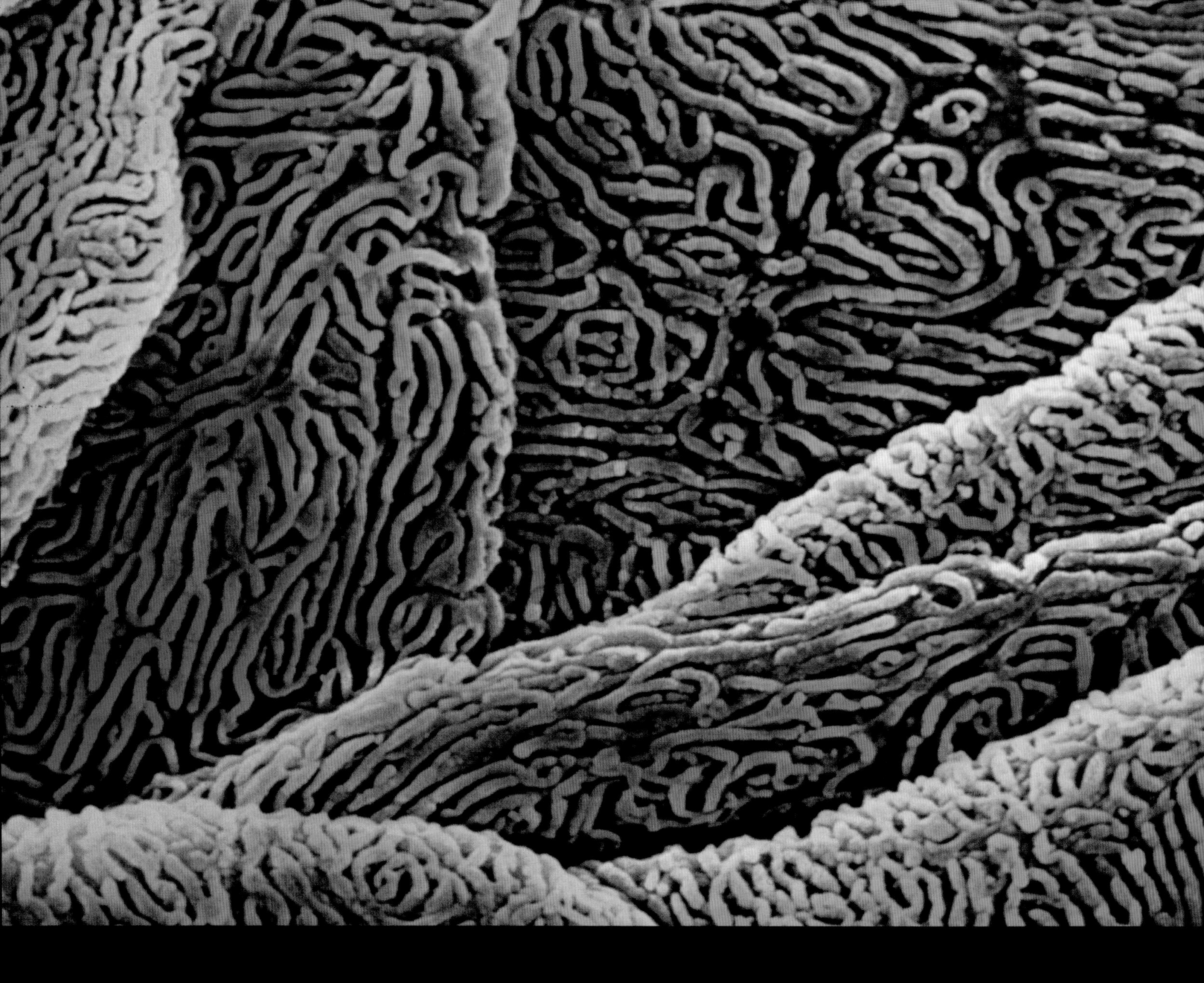

Above: Microplicae of the oesophagus (coloured scanning electron micrograph)

The oesophagus is the gullet, the tube which conveys food from the mouth to the stomach. When you swallow, the muscles of your gullet contract along its length, pushing the food down it. The surface of the gullet produces mucus which lubricates the lining of the gullet and assists the passage of food. Microplicae cover the surface, tiny ridges which act as dams to retain the mucus and prevent the oesophagus from drying out. (Magnification: x2600 at 6x7cm/2¼x2¾in size)

Right: Thyroid gland capillaries (coloured scanning electron micrograph)

This resin cast reveals capillaries in the thyroid gland, a butterfly-shaped organ at the base of the neck. The capillaries, the smallest blood vessels, are coiled around mound-like structures known as lobes. Within the lobes, clusters of round sacs called follicles produce hormones which control metabolism and growth. These hormones are distributed through the capillaries to almost every cell in the human body. They regulate the use of glucose for energy and help the skeleton retain calcium for bones. (Magnification: unknown)

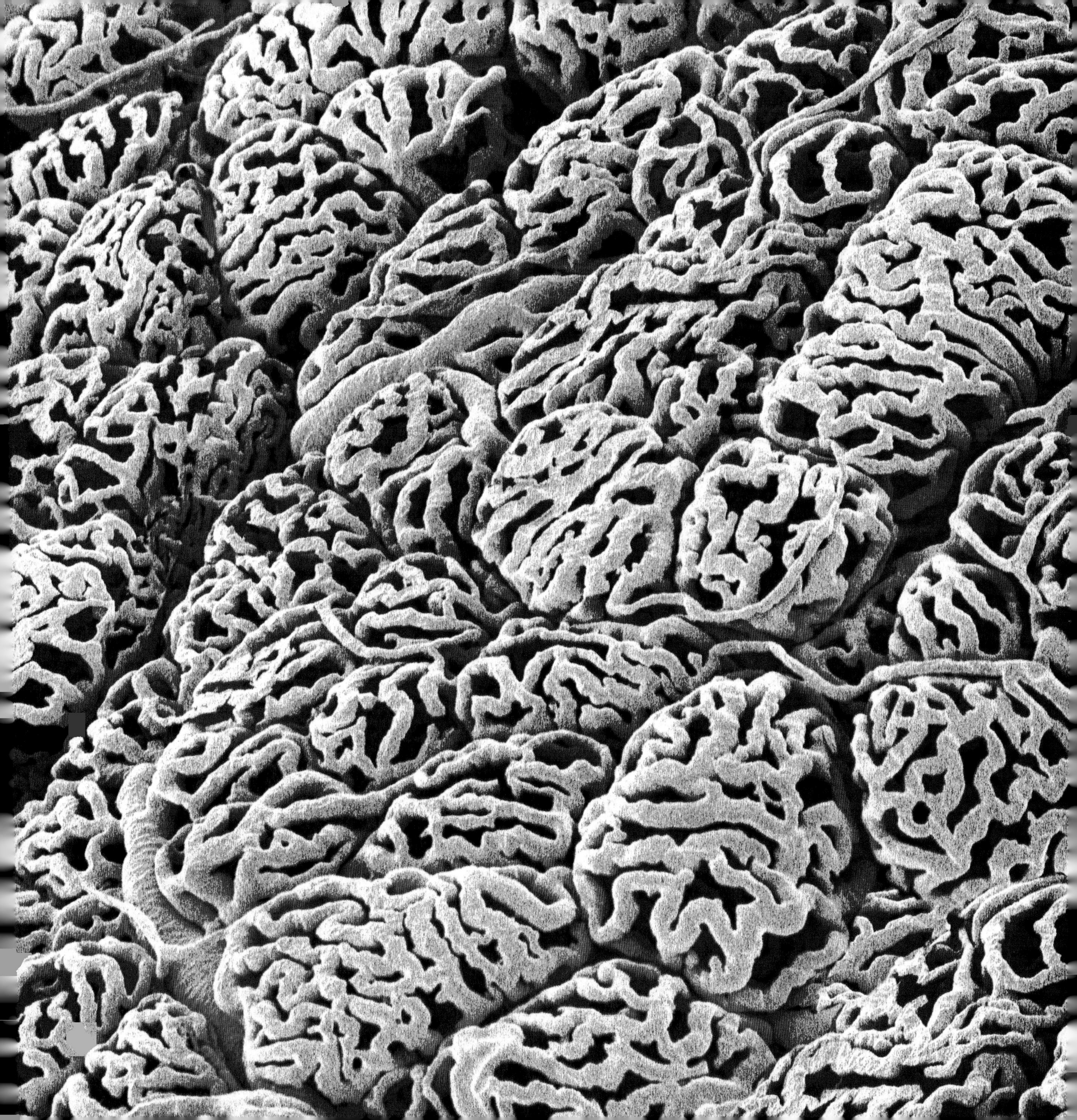

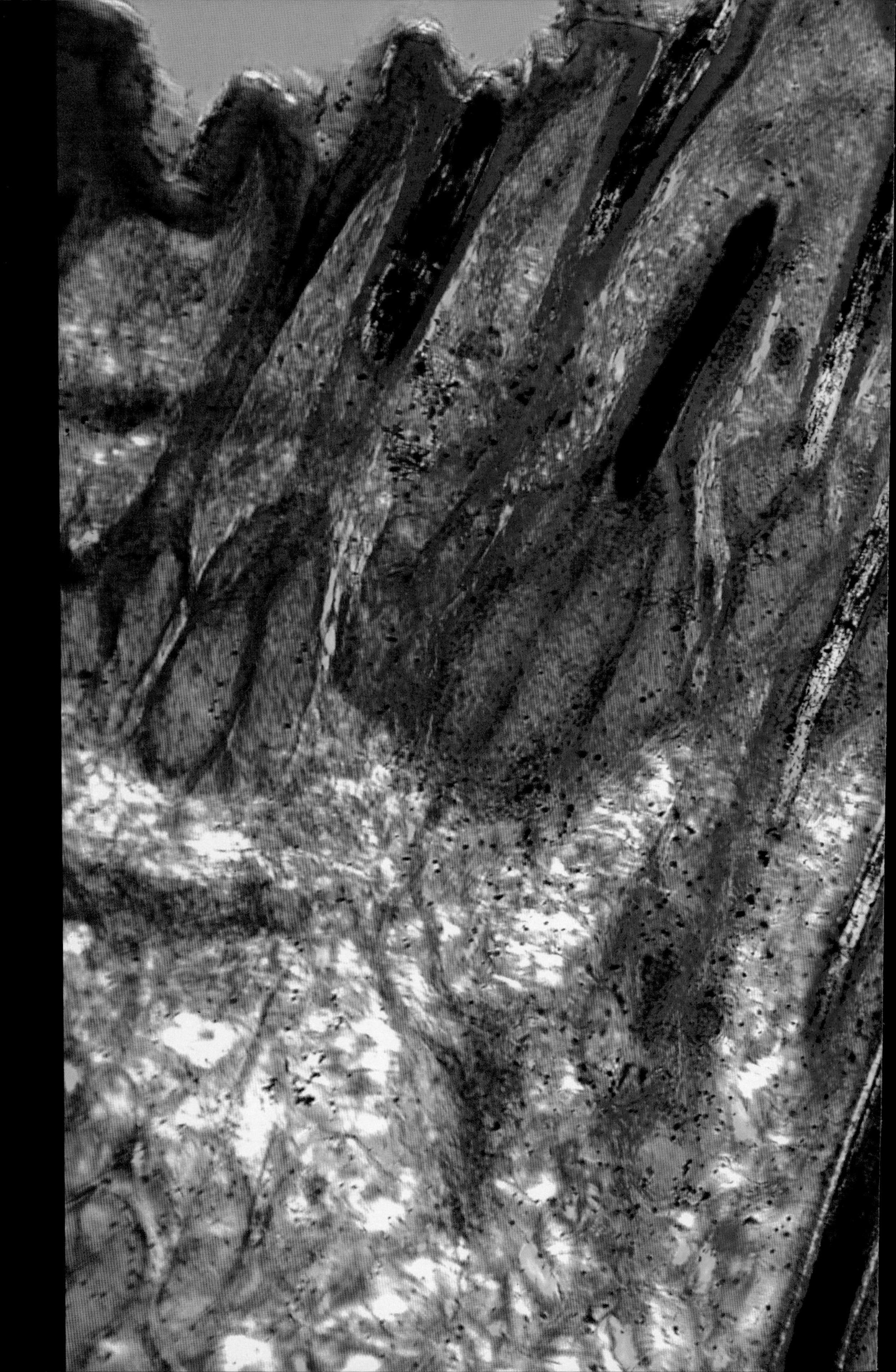

Human skin (polarized light micrograph)
The outer layer of the skin, the epidermis (top half of this image) consists of dead cells that are constantly sloughed off and replaced from below. These tightly packed cells contain high levels of a protein called keratin (yellow) which makes the skin waterproof and strong, to protect the organs inside. In this cross-section, you can also see hair follicles (black). Beneath the protective epidermis, and not seen in this image, is the dermis which contains blood vessels and nerves. (Magnification: unknown)

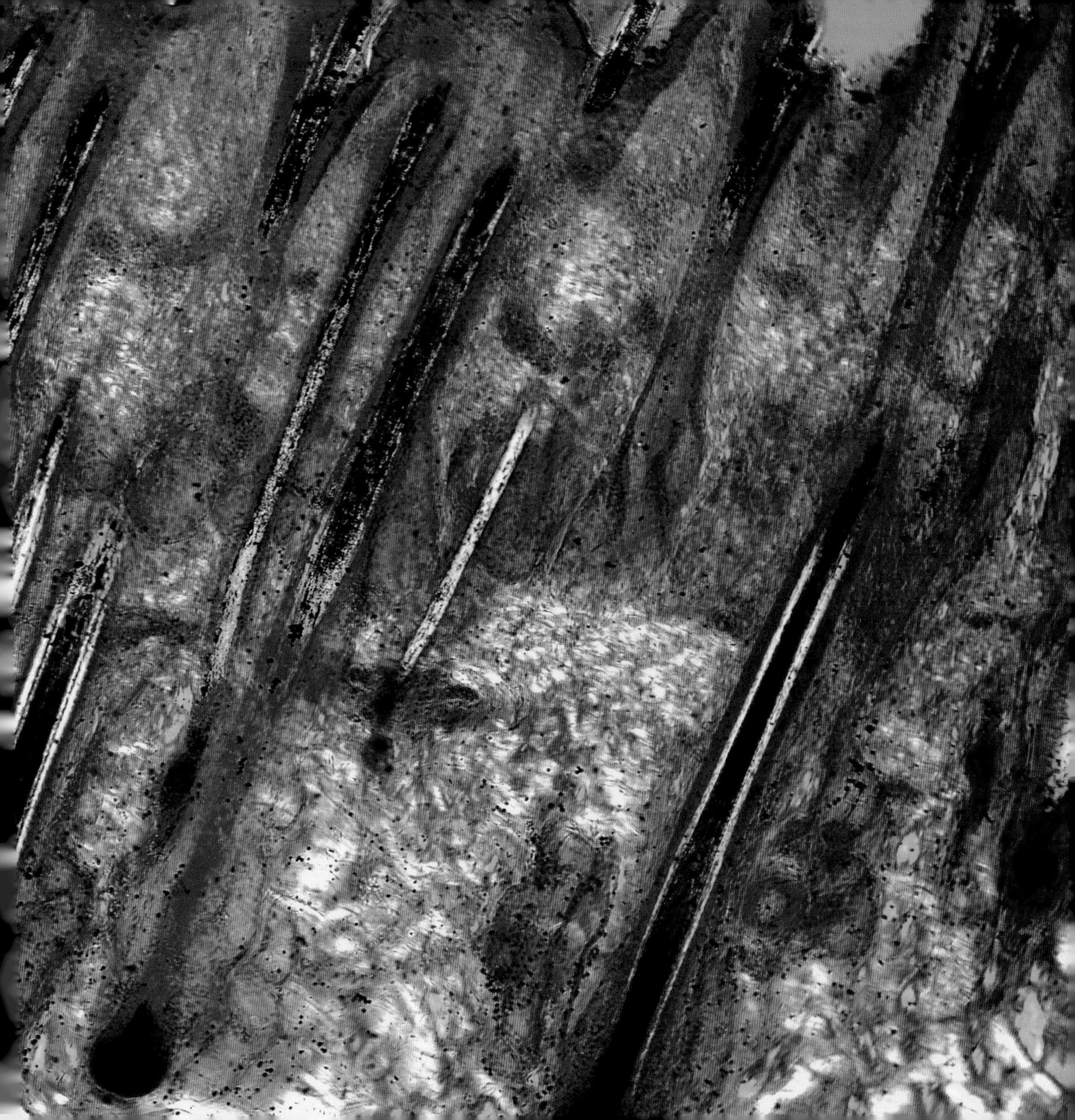

Melatonin crystals (polarized light micrograph)

Melatonin is produced by the pineal gland in the brain, the gland that controls the body's biological rhythm. The pineal gland receives messages relayed to it from the eye, and is less active in bright light. Melatonin is released at night and induces sleep. In middle age, melatonin secretion drops off. This may be responsible for aging symptoms such as insomnia and irritability. Pills containing melatonin can be taken to prevent jetlag. (Magnification: unknown)

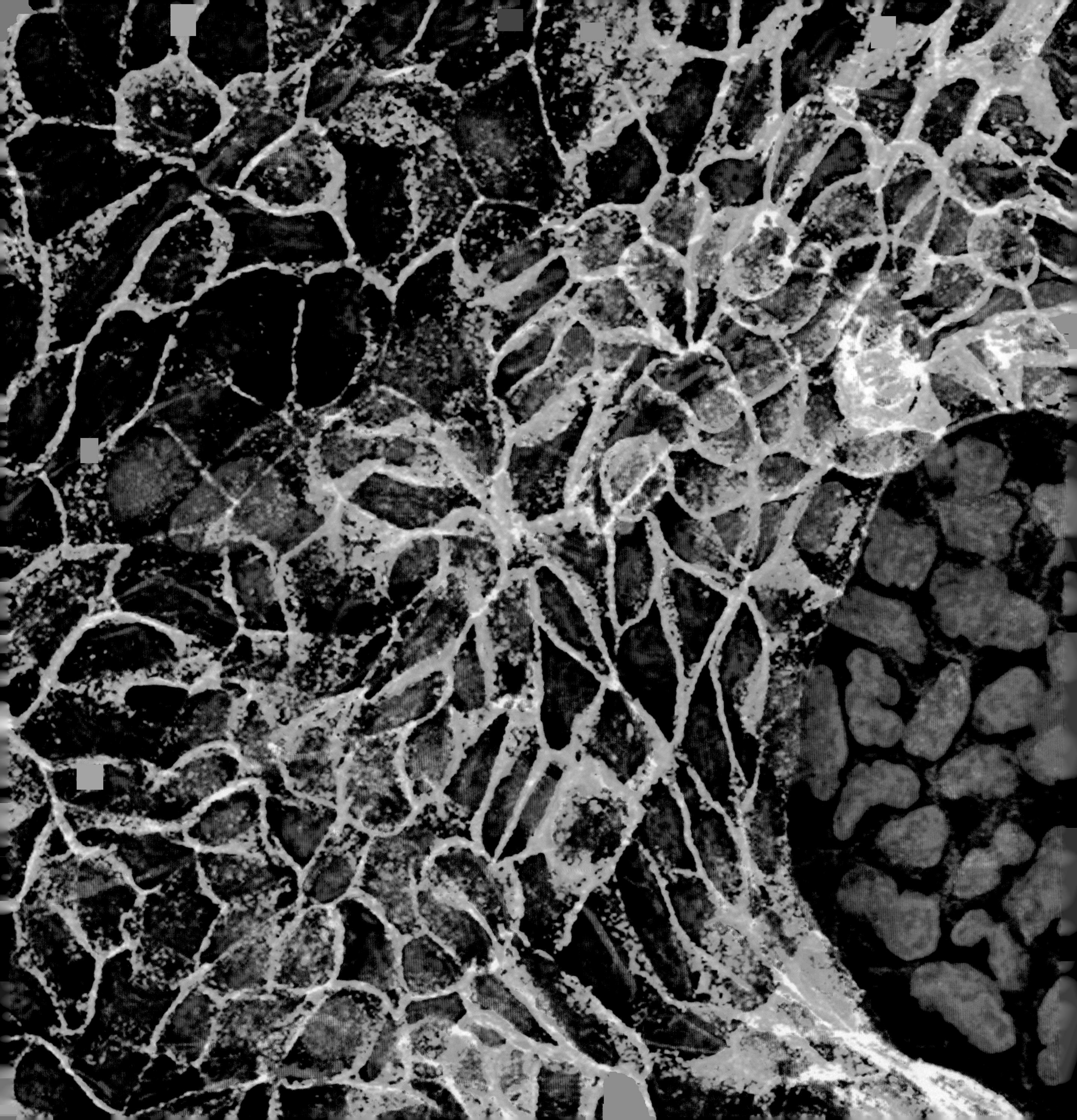

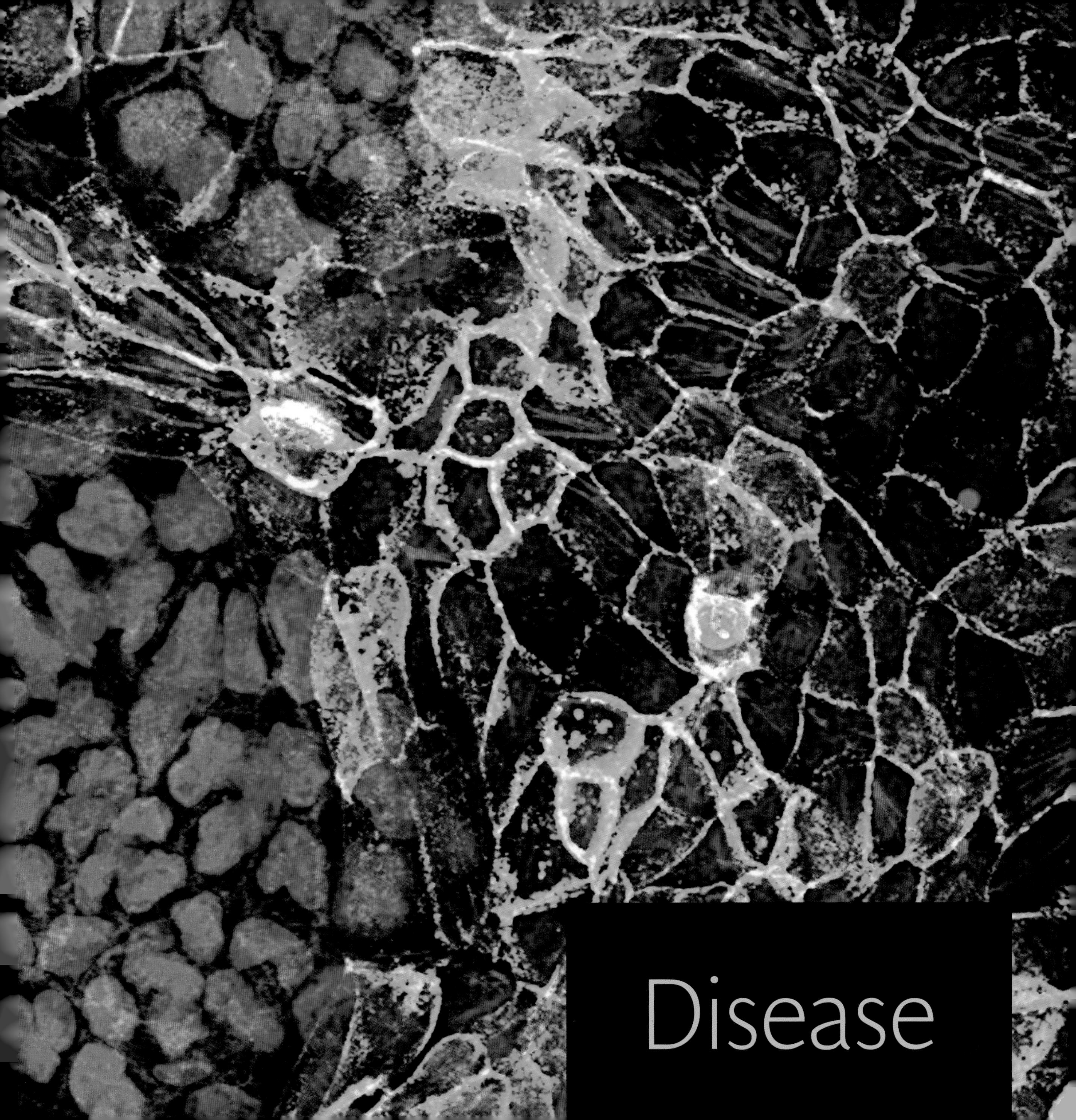

Disease

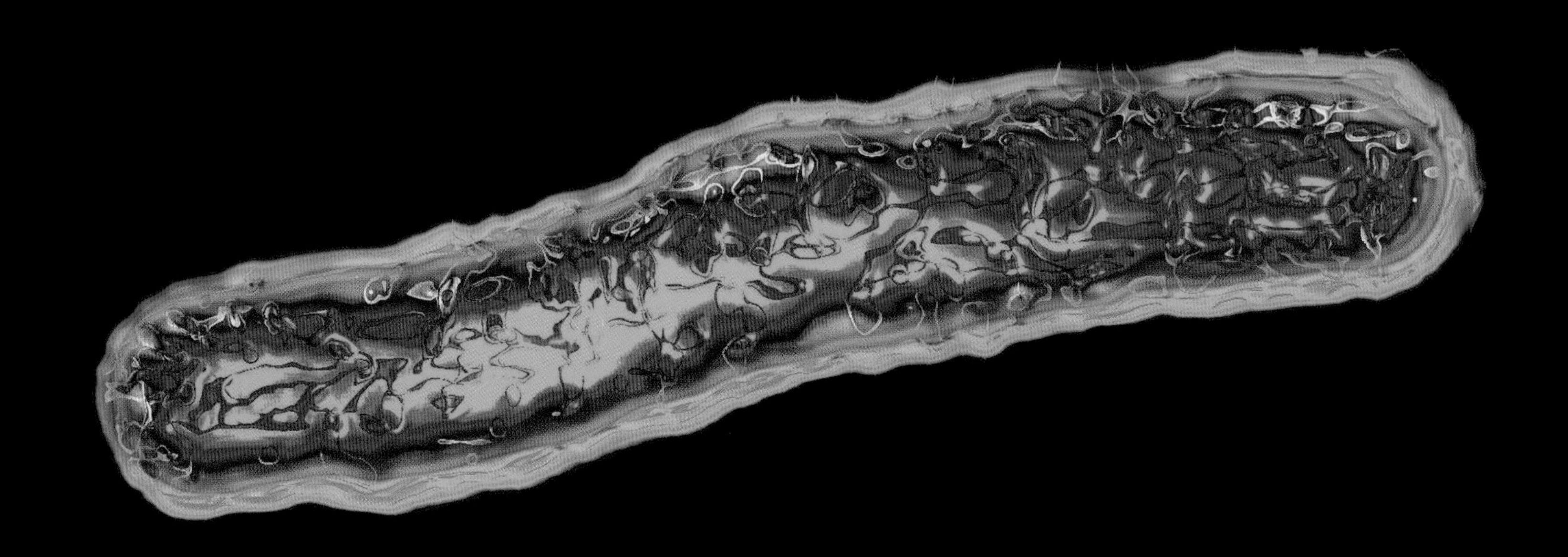

Previous page: Epithelial cancer cells (immunofluorescence light micrograph)
Epithelial cells form the surfaces which line cavities and cover structures of the body, as in the outer layer of skin, for example. This imaging technique uses fluorescent dyes to mark different proteins. Blue is DNA (deoxyribonucleic acid), the genetic material of cells. Green is E-Cadherin, a molecule which binds epithelial cells together. Normal cells have plenty of E-Cadherin, while the cancerous cells (centre) do not. In the background the structural fibres of the cells show up as red. (Magnification: unknown)

Above: *Mycobacterium tuberculosis* (coloured light micrograph)
This purple 'worm' is *Mycobacterium tuberculosis*, the pathogenic ('disease-spreading') bacterial species most commonly responsible for cases of tuberculosis. The disease typically infests the lungs, causing problems with breathing, and in extreme cases the rupture of the pulmonary artery which supplies blood to the lungs. The symptomatic cough helps to transmit the bacteria to other people; hence the saying 'coughs and sneezes spread diseases.' Tuberculosis kills almost two million people each year worldwide. (Magnification: x50000 at 10cm/4in size)

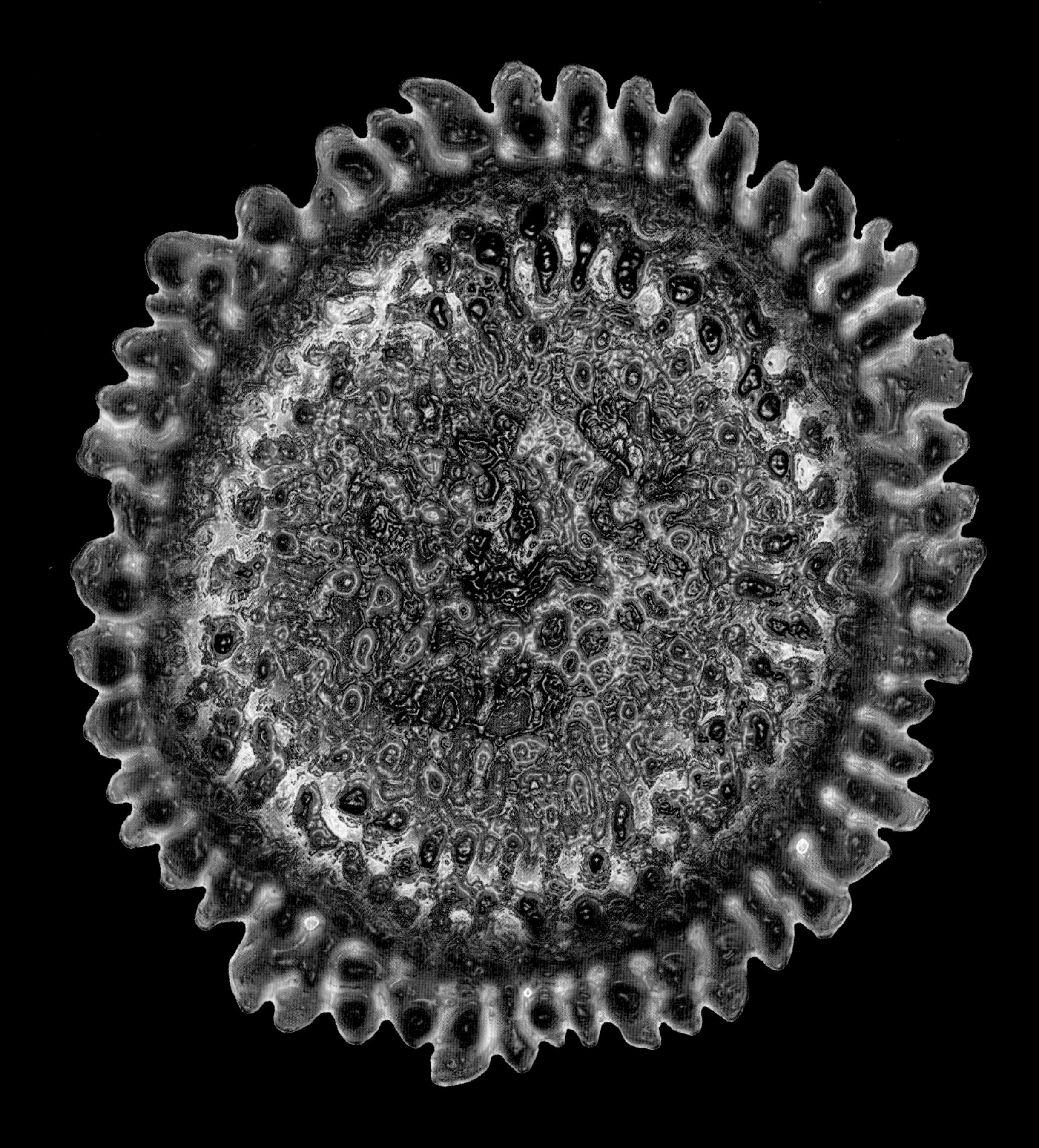

Avian influenza virus, H5N1
(coloured and filtered transmission electron micrograph)

The H5N1 virus causes a serious and contagious respiratory disease in humans, commonly known as bird flu. It is caught from infected birds, amongst which (particularly in southeast Asia) it is widespread. There are very rare instances of it being passed form human to human. It is a virus of the Orthomyxoviridae family, which includes viruses responsible for the deadly Spanish flu outbreak of 1918, Hong Kong flu in 1968, and the Swine flu epidemic of 2009. (Magnification: unknown)

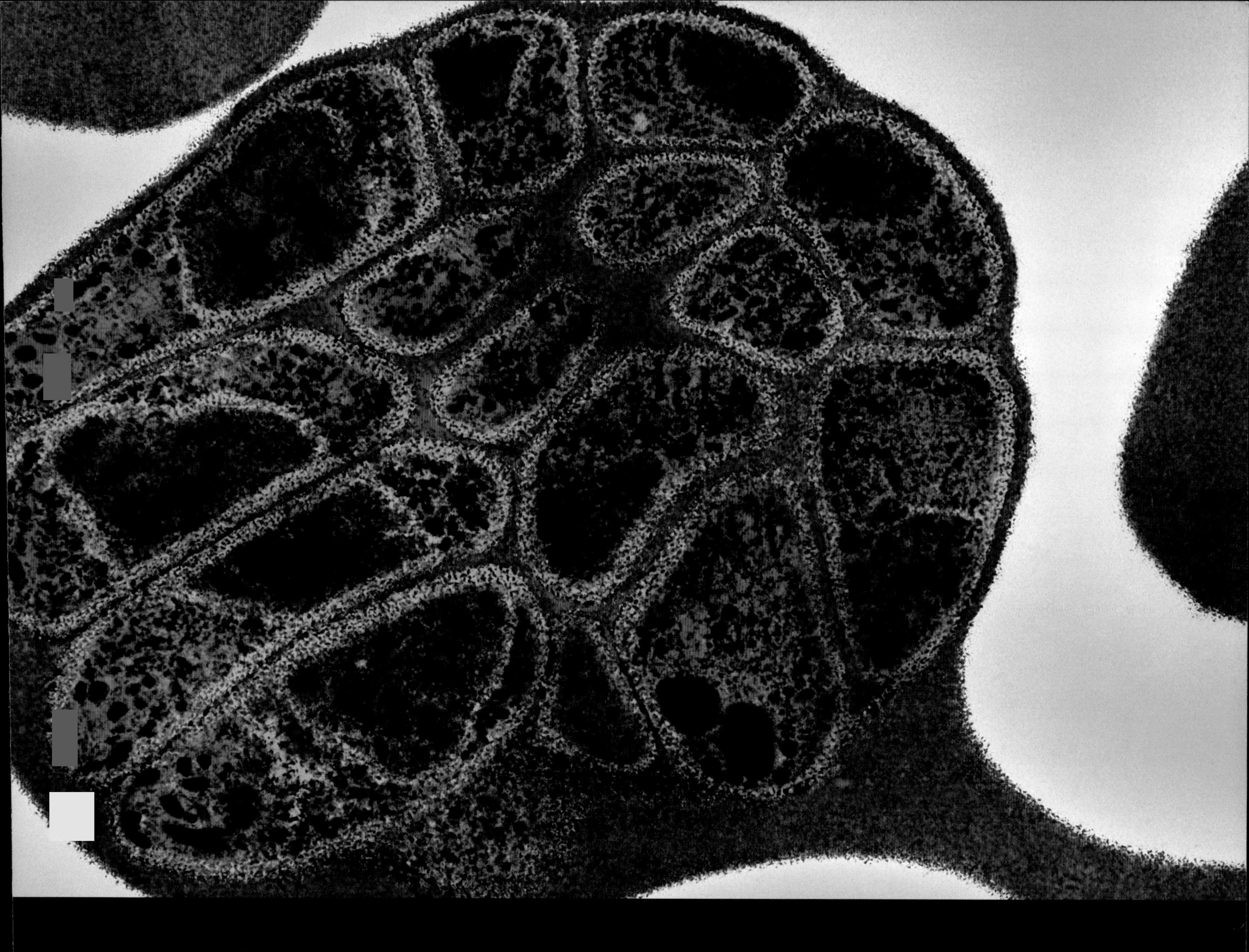

Left: Macrophage engulfing TB bacteria (coloured scanning electron micrograph)
Macrophages are cells of the body's immune system. They destroy pathogens, dead cells and cellular debris in a process called phagocytosis, by engulfing their prey. Here a normally round macrophage (red) has developed extending folds called pseudopods, to embrace a tuberculosis bacterium (*Mycobacterium tuberculosis*, yellow). It will absorb the bacterium into its body where it will break it down into harmless waste. (Magnification: x750 at 10cm/4in size)

Above: Malaria (coloured transmission electron micrograph)
A human red blood cell (red) infected with the malarial parasite *Plasmodium* (blue). The single-celled *Plasmodium* parasites are seen in the swollen region of the cell. Malaria is spread to humans by species of tropical mosquitoes (*Anopheles*). Infection spreads from the liver to the blood, where the *Plasmodium* multiplies inside red blood cells. Mature parasites break out of blood cells to cause further infection, producing bouts of shivering fever and sweating that may be fatal. (Magnification: x13000 at 5x7cm/2x2¾in size)

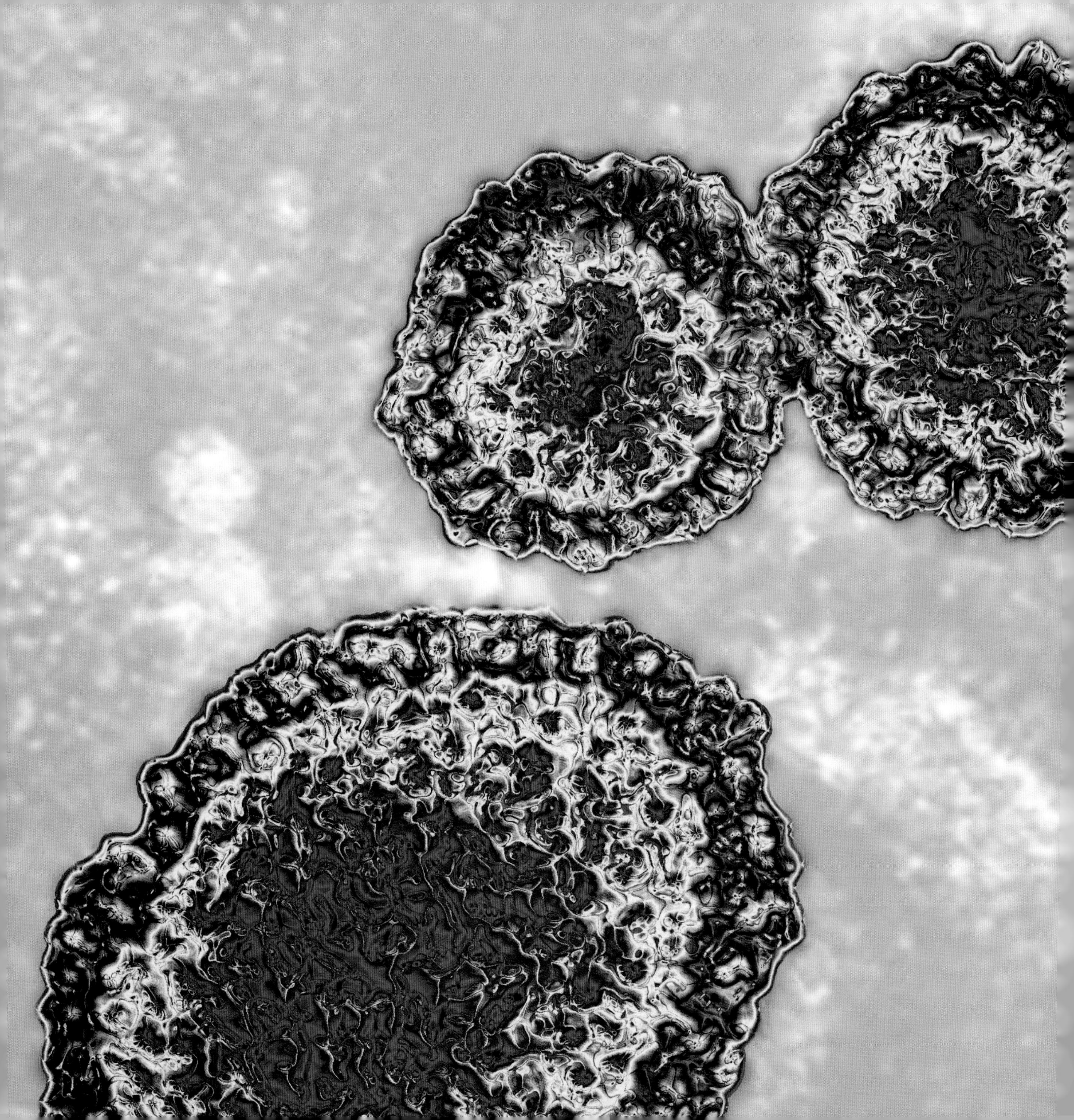

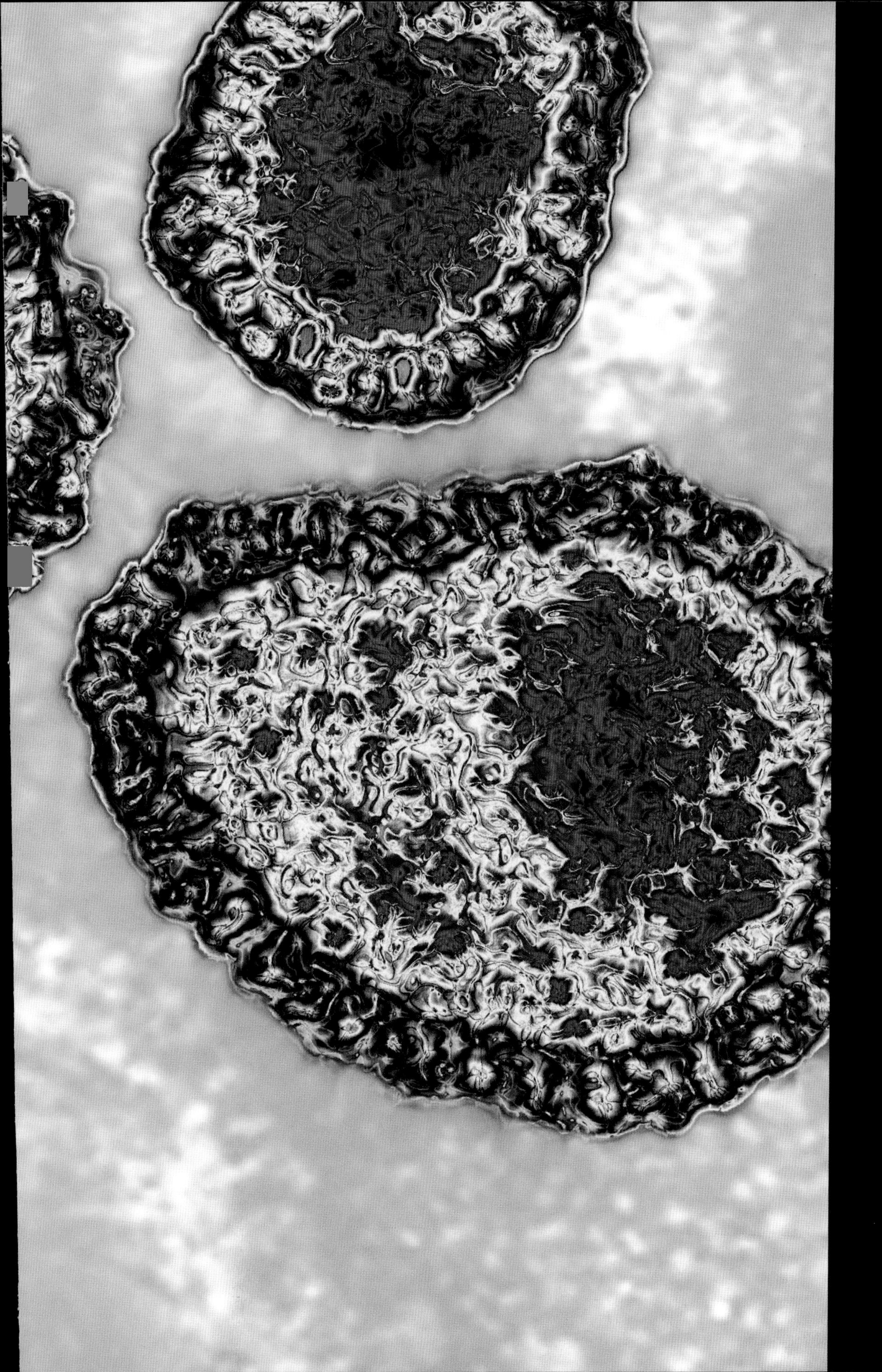

Influenza A H1N1 virus particles (coloured transmission electron micrograph)
Influenza A viruses can infect humans, pigs, birds and horses. The H1N1 strain caused the Swine flu outbreak of 2009. At the centre of each virus is its genetic fingerprint (the ribonucleic acid, pink), surrounded by a protective protein shell (the nucleocapsid, yellow). The enclosing fatty envelope (green) contains two types of protein, haemagglutinin and neuraminidase (the 'H' and 'N' in the strain's codename), the levels of which determine the strain of virus. (Magnification: unknown)

Gallstone crystals (coloured scanning electron micrograph)

Gallstones mostly consist of cholesterol, but can also contain calcium and bilirubin (a product of old red blood cells). They form in the gallbladder (from which bile is released into the small intestine) when there is an imbalance in the chemical composition of the bile. Gallstones are usually symptomless, unless one obstructs the bile duct. In that case they cause acute pain, jaundice and infection. Treatment is with drugs to dissolve the stones or surgical removal of the gallbladder. (Magnification: x750 at 10cm/4in size)

Vaginal cancer cells (coloured scanning electron micrograph)

Vaginal cancer is most common in women over 60, and the most common form of it is the squamous cell carcinoma seen here. Squamous cells are flat, many-sided, scale-like cells, part of the epithelium – the outermost layer that lines cavities and covers surfaces in the body. Skin is a form of epithelium, and squamous cell carcinoma is one of the most frequently found skin cancers. The many microvilli (small projections) on the cell's surface are typical of cancer cells. (Magnification: x1000 at 10cm/4in size)

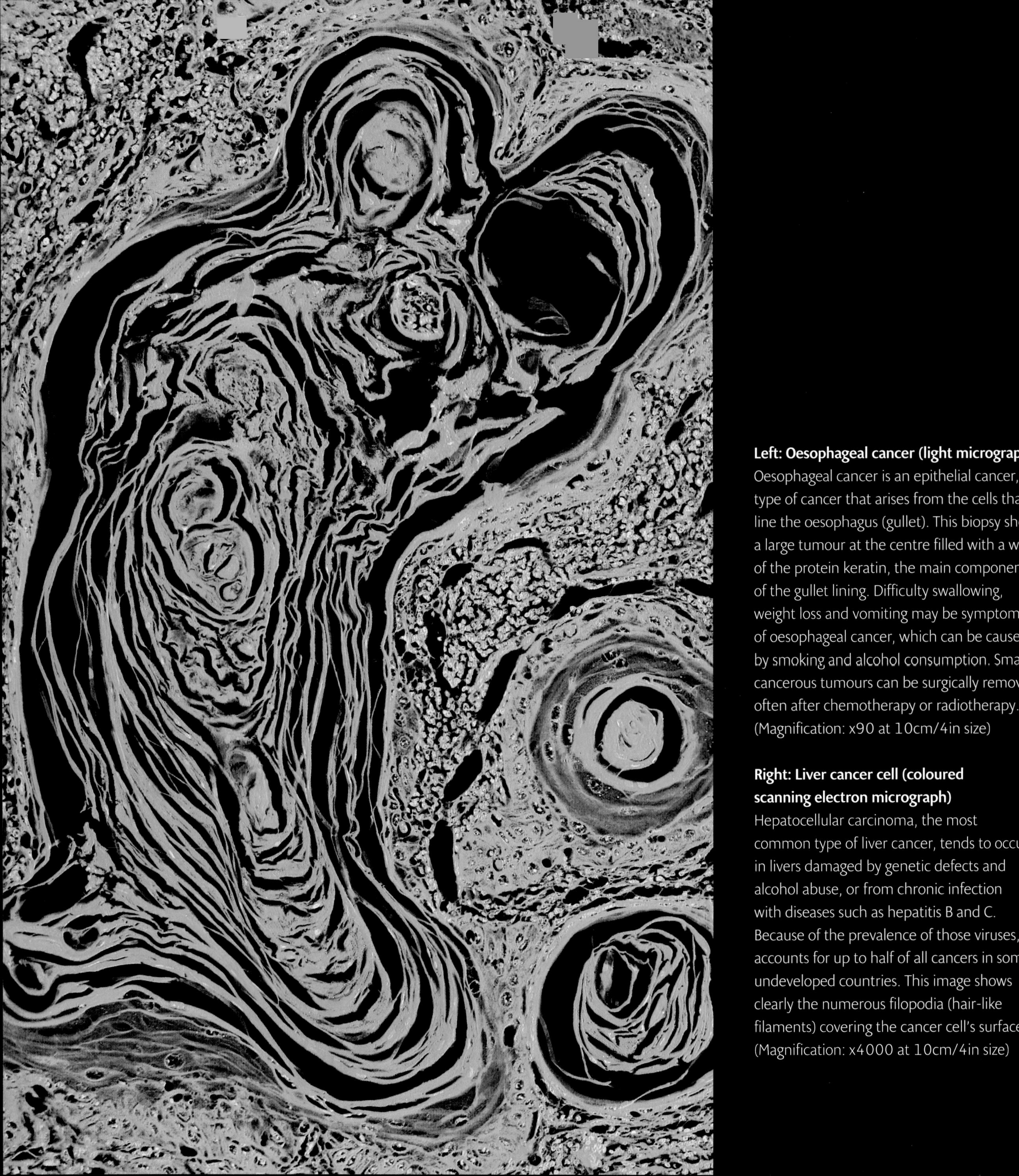

Left: Oesophageal cancer (light micrograph)
Oesophageal cancer is an epithelial cancer, a type of cancer that arises from the cells that line the oesophagus (gullet). This biopsy shows a large tumour at the centre filled with a whorl of the protein keratin, the main component of the gullet lining. Difficulty swallowing, weight loss and vomiting may be symptoms of oesophageal cancer, which can be caused by smoking and alcohol consumption. Small cancerous tumours can be surgically removed, often after chemotherapy or radiotherapy. (Magnification: x90 at 10cm/4in size)

Right: Liver cancer cell (coloured scanning electron micrograph)
Hepatocellular carcinoma, the most common type of liver cancer, tends to occur in livers damaged by genetic defects and alcohol abuse, or from chronic infection with diseases such as hepatitis B and C. Because of the prevalence of those viruses, it accounts for up to half of all cancers in some undeveloped countries. This image shows clearly the numerous filopodia (hair-like filaments) covering the cancer cell's surface. (Magnification: x4000 at 10cm/4in size)

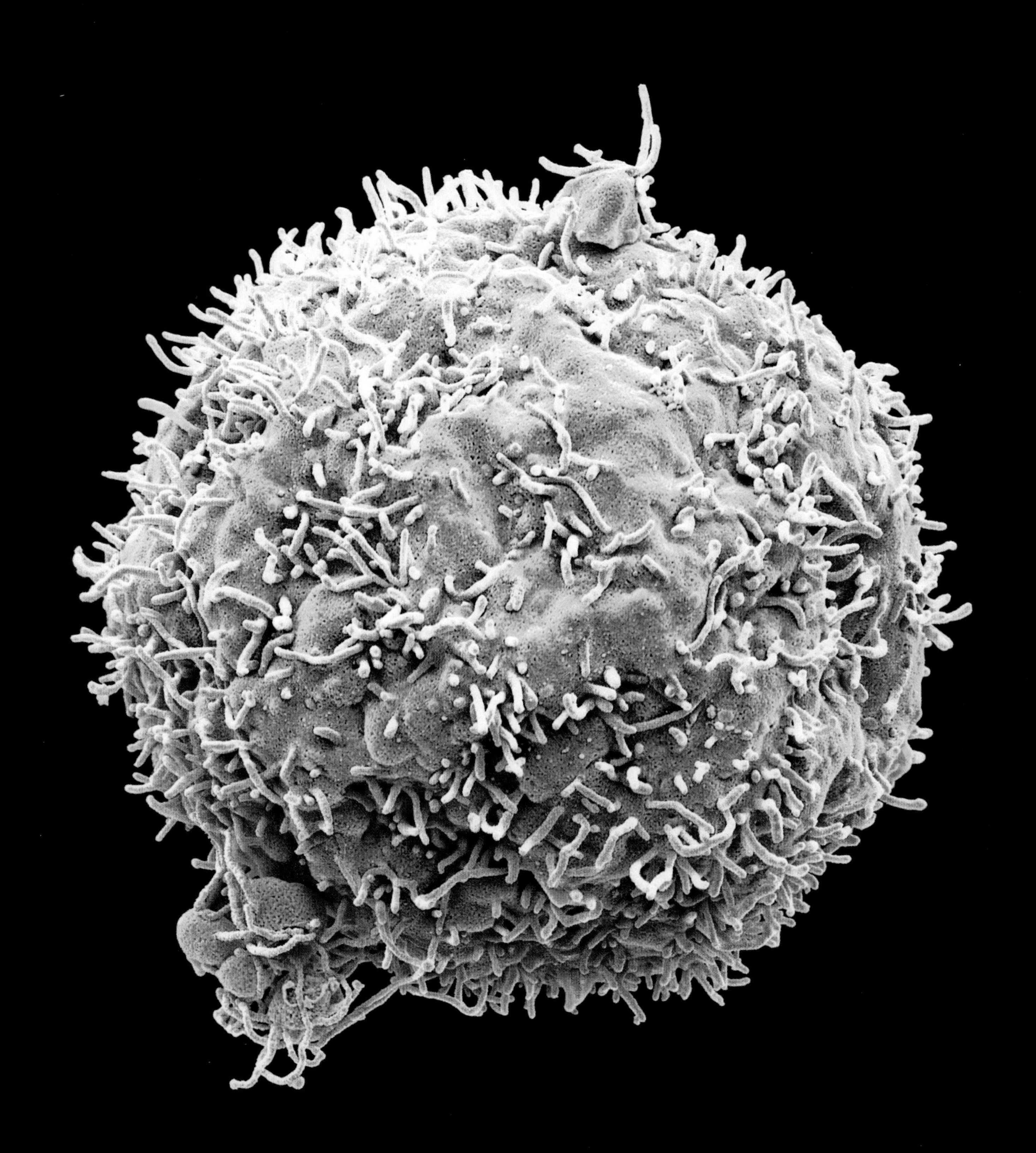

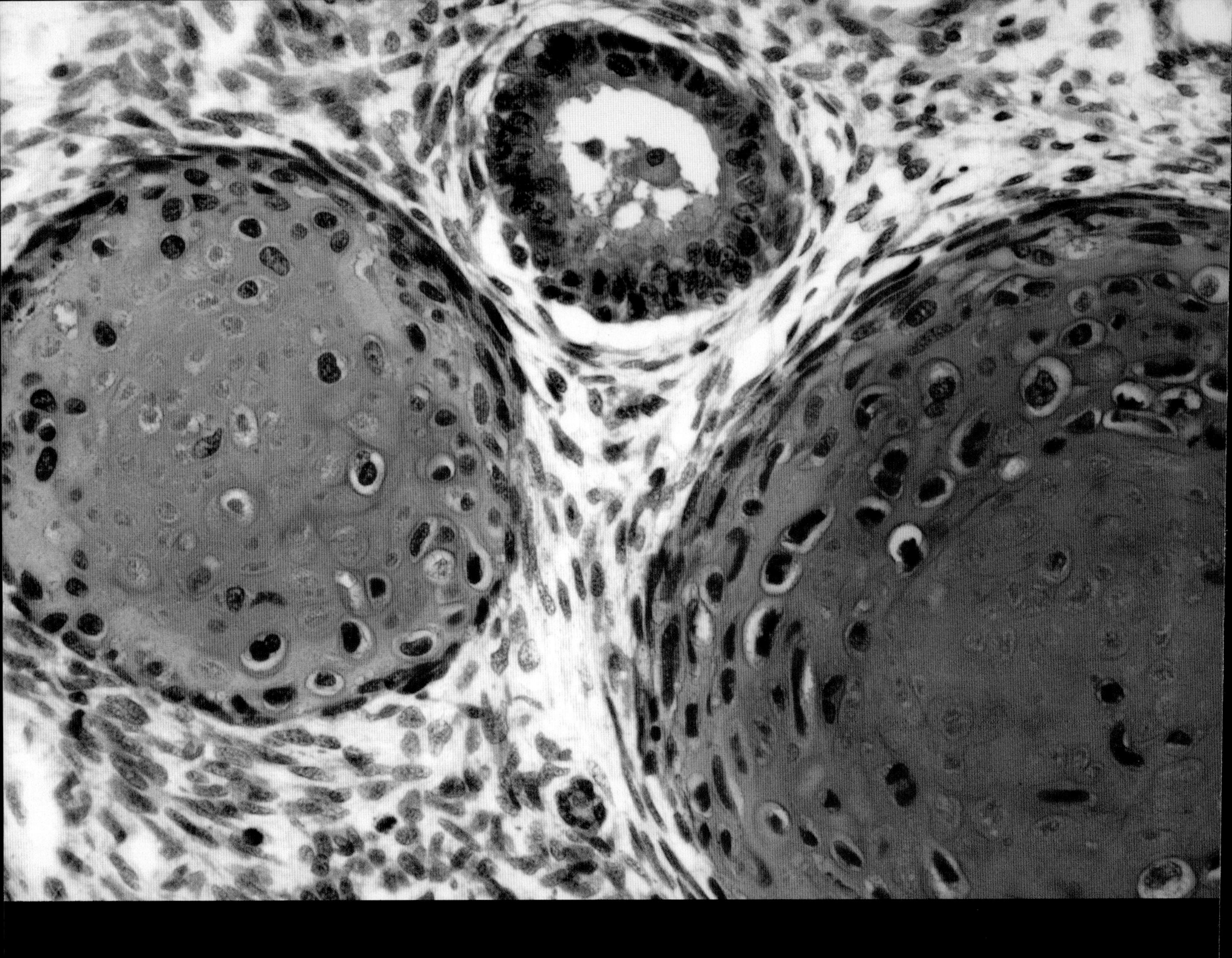

Above: Testicular cancer (light micrograph)
A cross-section of a teratoma of the testis, a rare type of cancer. Teratomas are enclosed tumours found throughout the body that form around tissue or organ components, such as hair or teeth. They are thought to be present at birth and are usually harmless; but if they occur in the testicles they are sometimes more aggressively malignant. When present, they feel like small, usually painless, lumps on the testis. (Magnification: x200 at 10cm/4in size)

Right: Breast cancer (light micrograph)
The most common form of breast cancer is the one seen here, a ductal carcinoma (pink) that arises from the lining of the milk ducts. The stroma (yellow), the breast tissue surrounding the cancer cells, is visibly fibrous. It is infiltrated with white blood cells – lymphocytes and plasma cells – which are part of the body's immune system. High levels of tumour-infiltrating lymphocytes (TILs) are often a sign of a better prognosis for the treatment of cancer. (Magnification: x200 at 10cm/4in size)

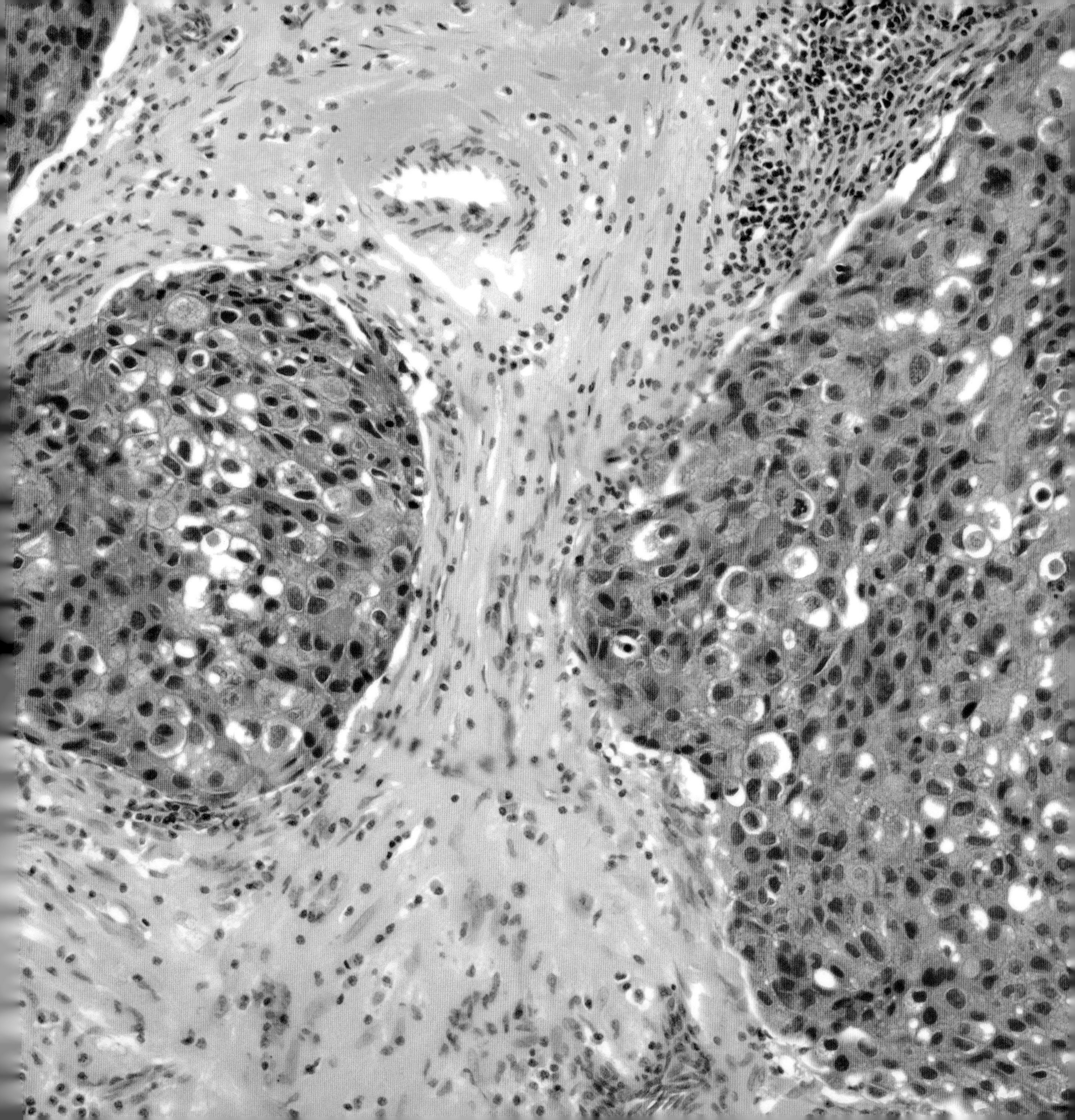

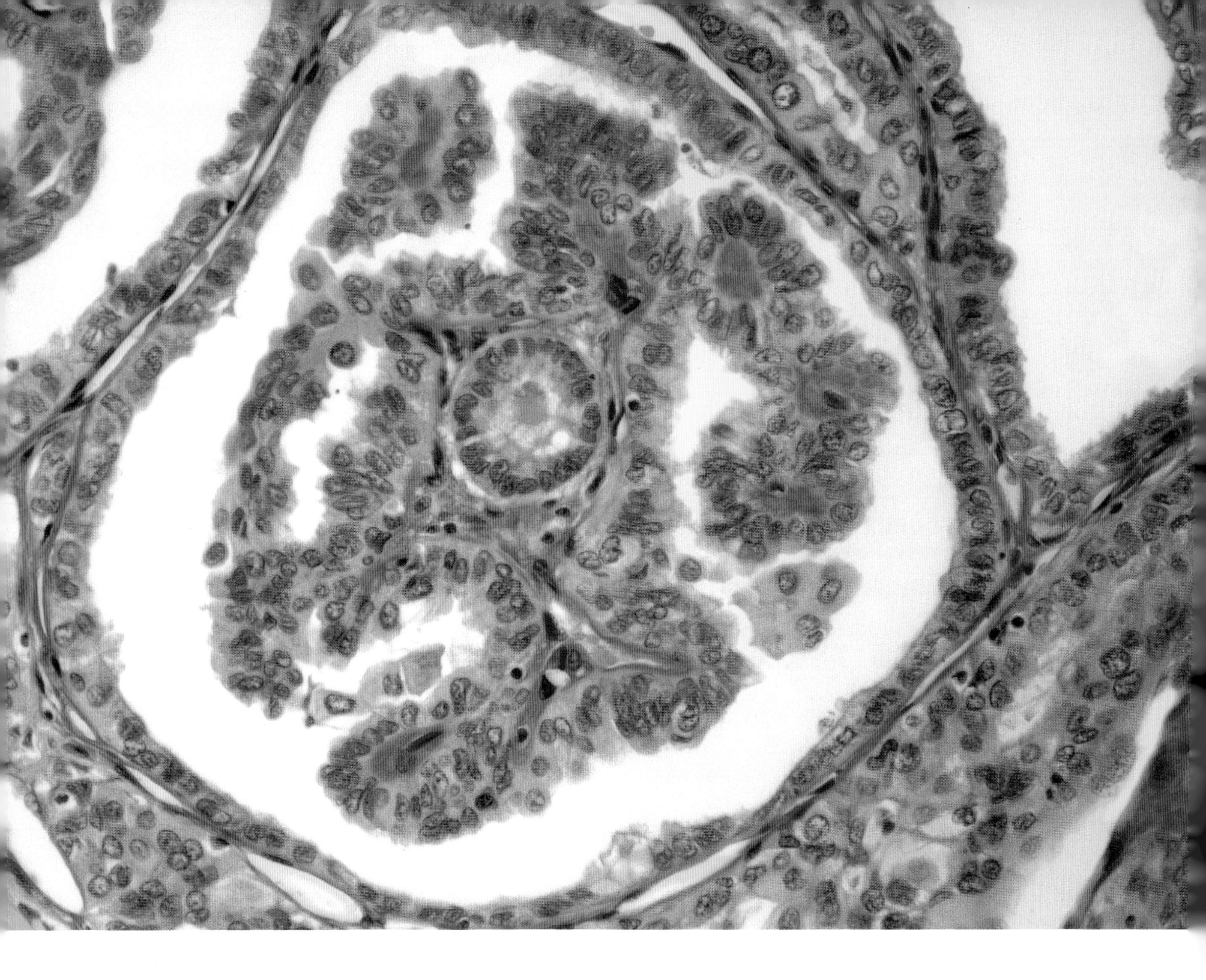

Thyroid cancer (light micrograph)

Thyroid cancer is rare. The most common form, in three out of four diagnoses, is papillary adenocarcinoma, seen here in cross-section. Within the sac of a thyroid follicle (which produces hormones), the walls (pink) are lined with hormone-generating cells (yellow). When the cells are cancerous, the nuclei (green) tend to overlap, and they appear 'empty' and uniformly coloured. Thyroid cancer develops slowly and has good survival rates after treatment. (Magnification: x200 at 10cm/4in size)

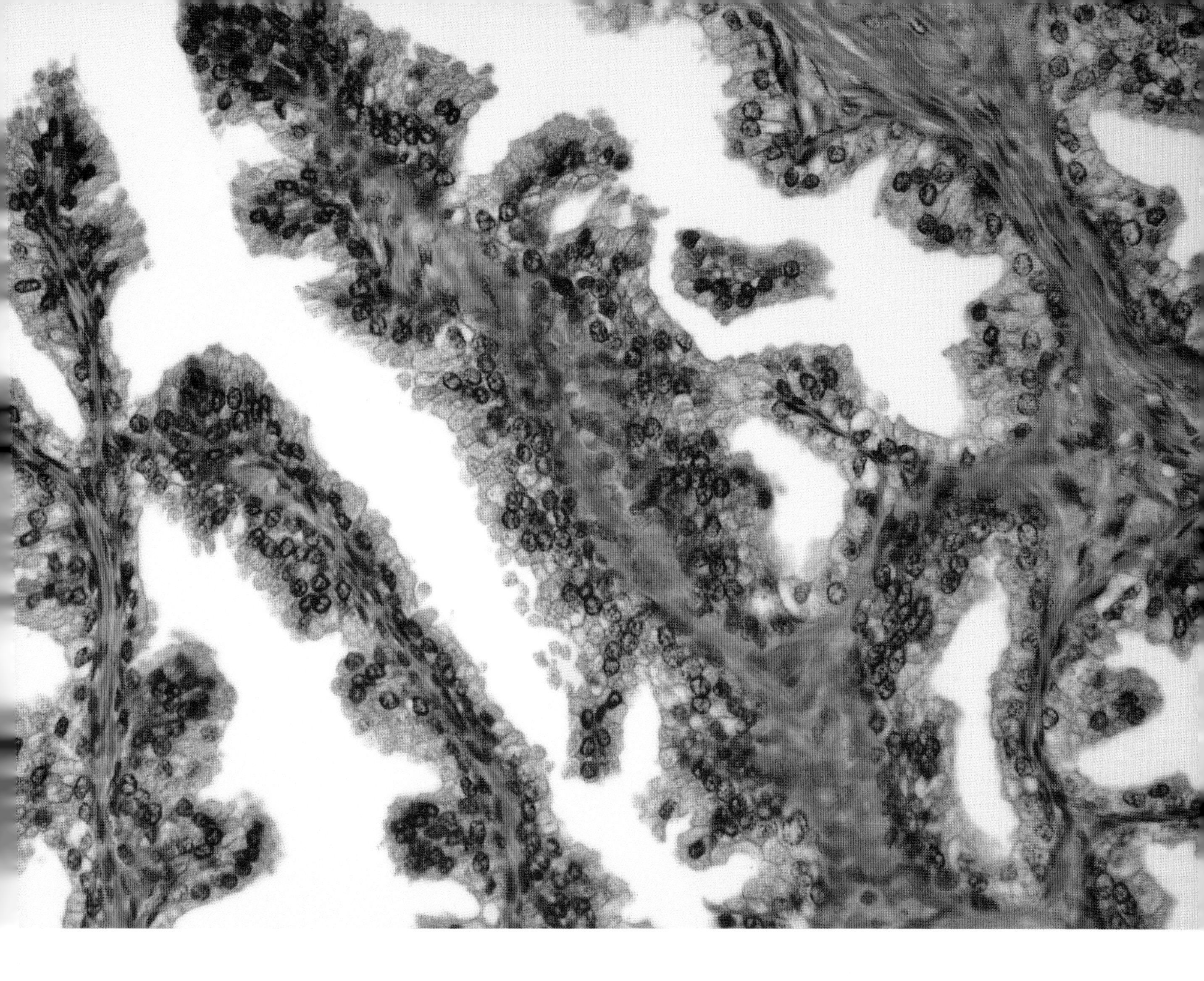

Prostate cancer (light micrograph)
The prostate secrets a milky fluid which makes up around half the volume of semen and contributes to the motility of sperm. Most prostate cancers develop slowly and remain undetected. Some however are more aggressive, and in Britain prostate cancer is the most frequent cause of death amongst male cancer patients. This cross-section shows the typically enlarged nuclei (purple) of cells within an affected prostate. (Magnification: x200 at 10cm/4in size)

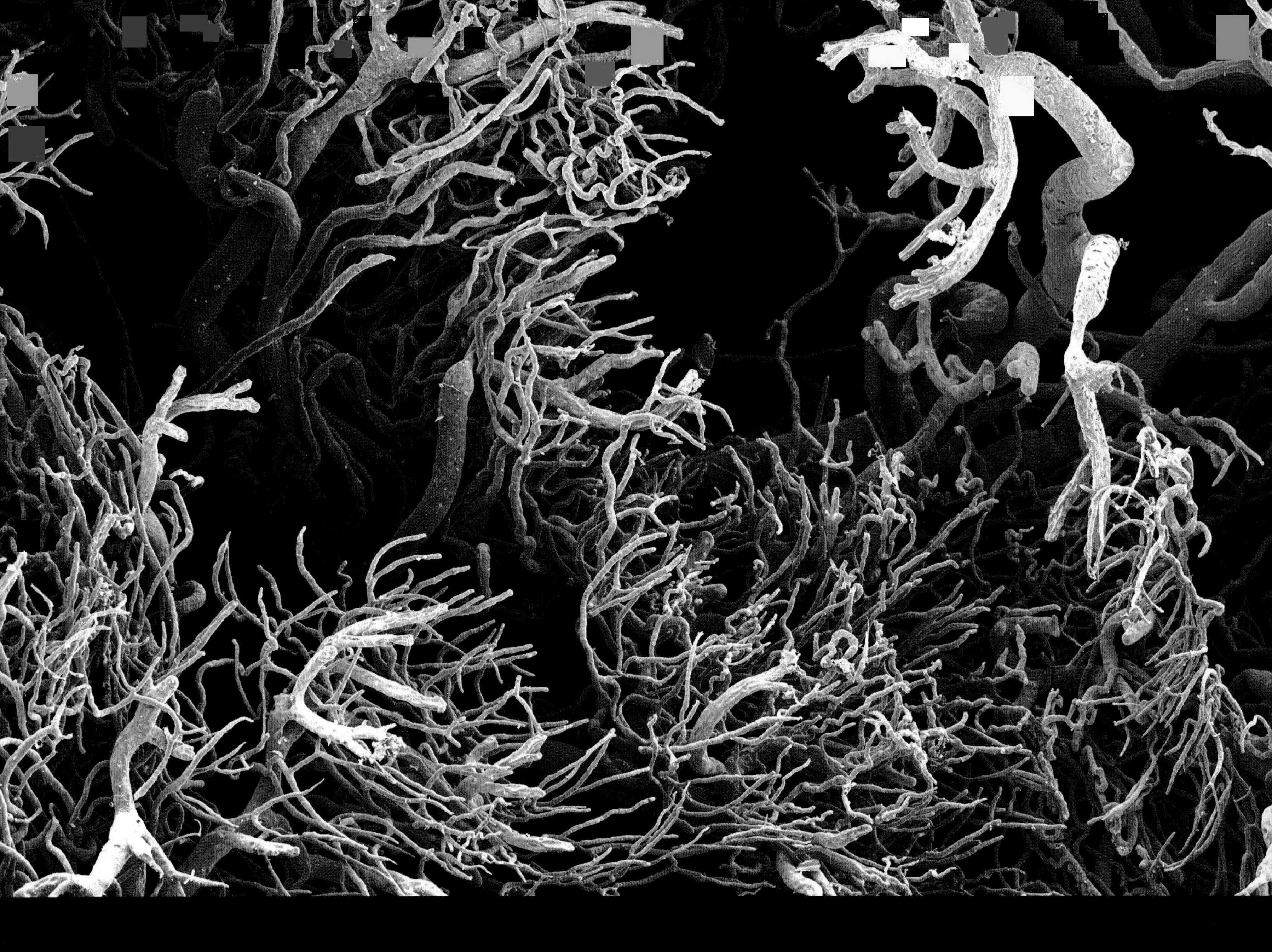

Tumour blood vessels (coloured scanning electron micrograph)
This is a resin cast of the branching network of vessels surrounding an intestinal tumour, supplying it with blood. Tumours are the uncontrolled growth of a tissue, and they trigger a growth of new blood vessels to supply the blood for such growth. The blood vessels form a chaotic structure here, as opposed to the normally ordered structure found in the intestines. Tumours may be malignant, invading and destroying other tissue as a cancer. (Magnification: unknown)

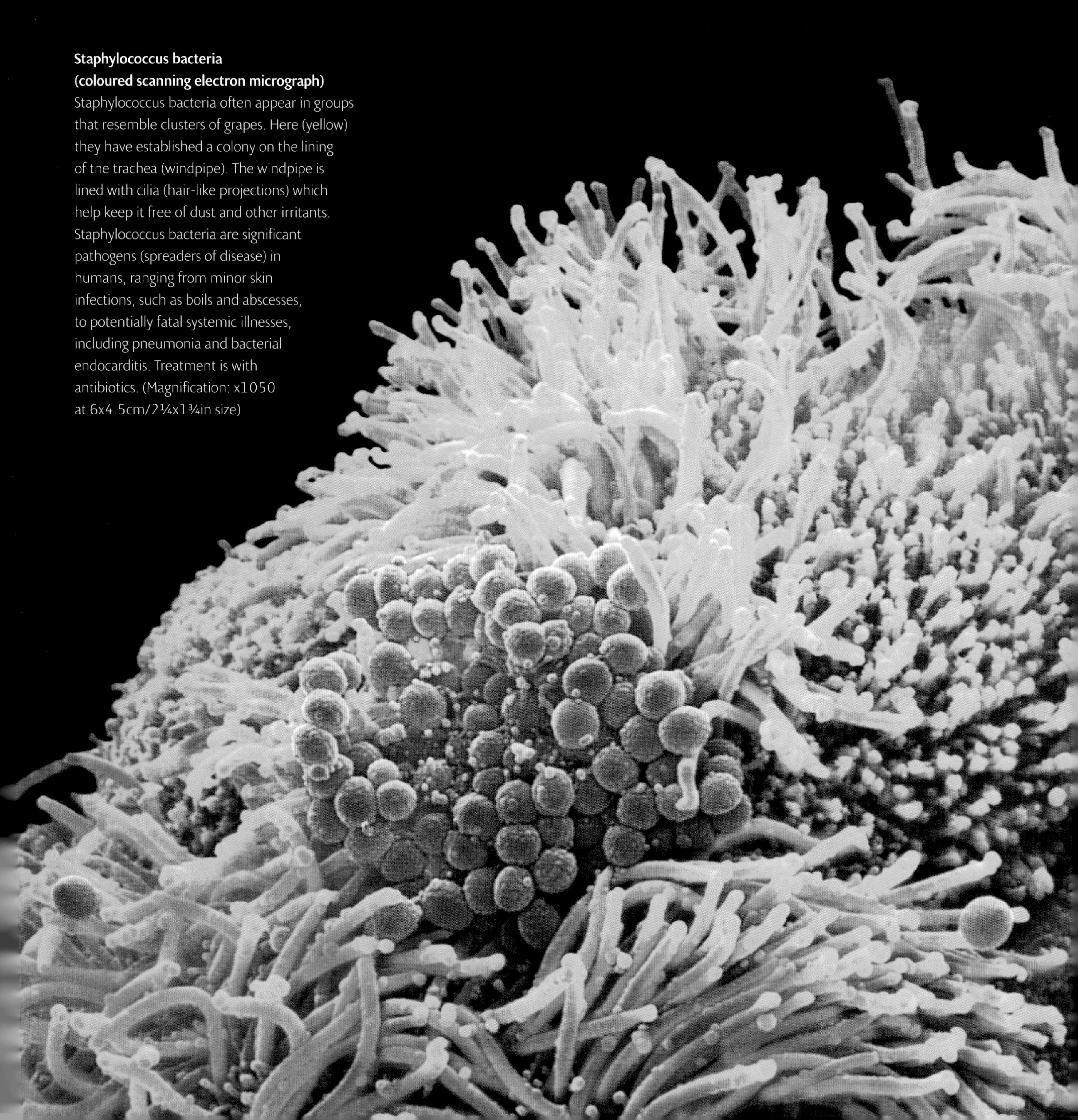

Staphylococcus bacteria (coloured scanning electron micrograph)
Staphylococcus bacteria often appear in groups that resemble clusters of grapes. Here (yellow) they have established a colony on the lining of the trachea (windpipe). The windpipe is lined with cilia (hair-like projections) which help keep it free of dust and other irritants. Staphylococcus bacteria are significant pathogens (spreaders of disease) in humans, ranging from minor skin infections, such as boils and abscesses, to potentially fatal systemic illnesses, including pneumonia and bacterial endocarditis. Treatment is with antibiotics. (Magnification: x1050 at 6x4.5cm/2¼x1¾in size)

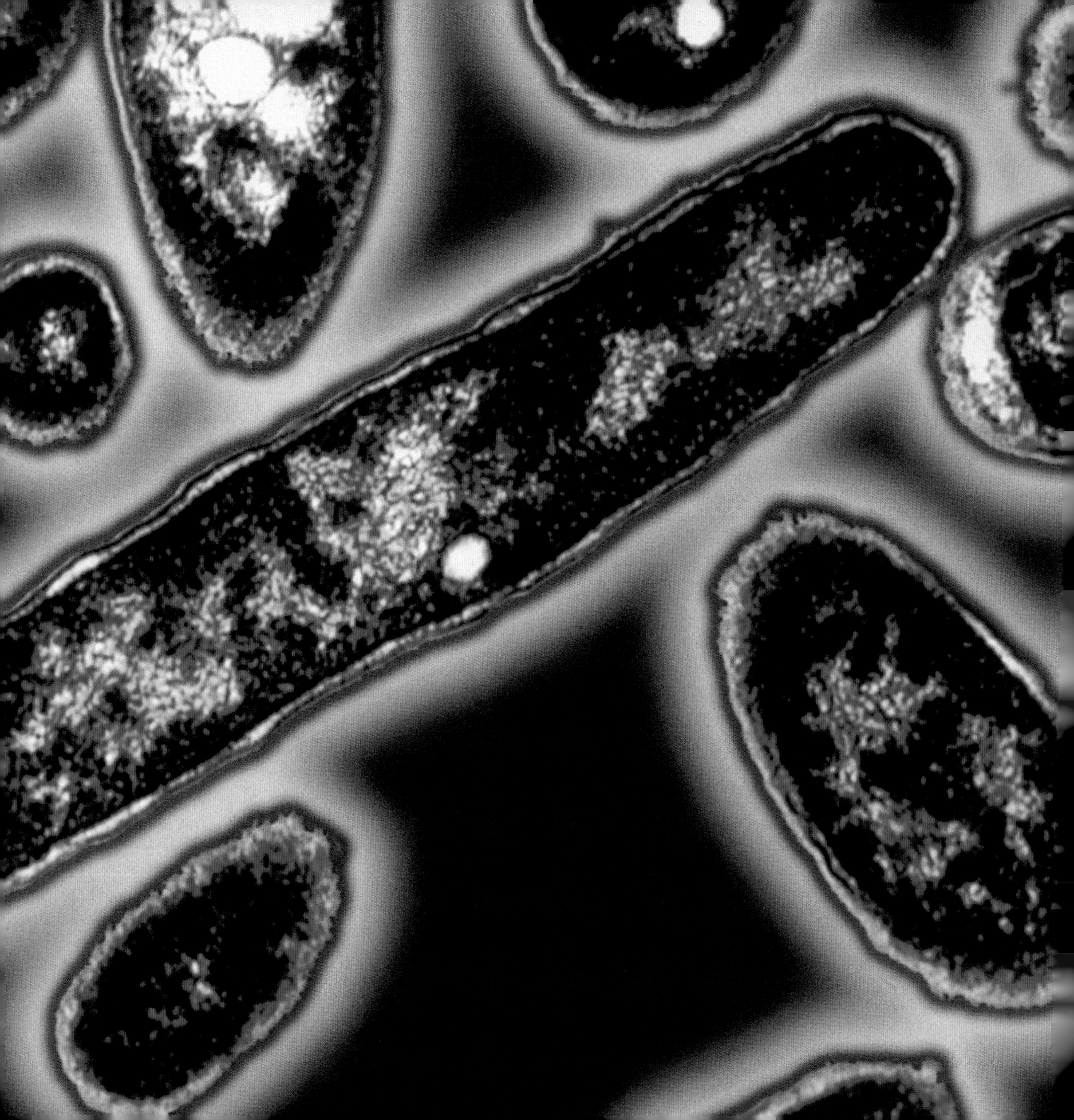

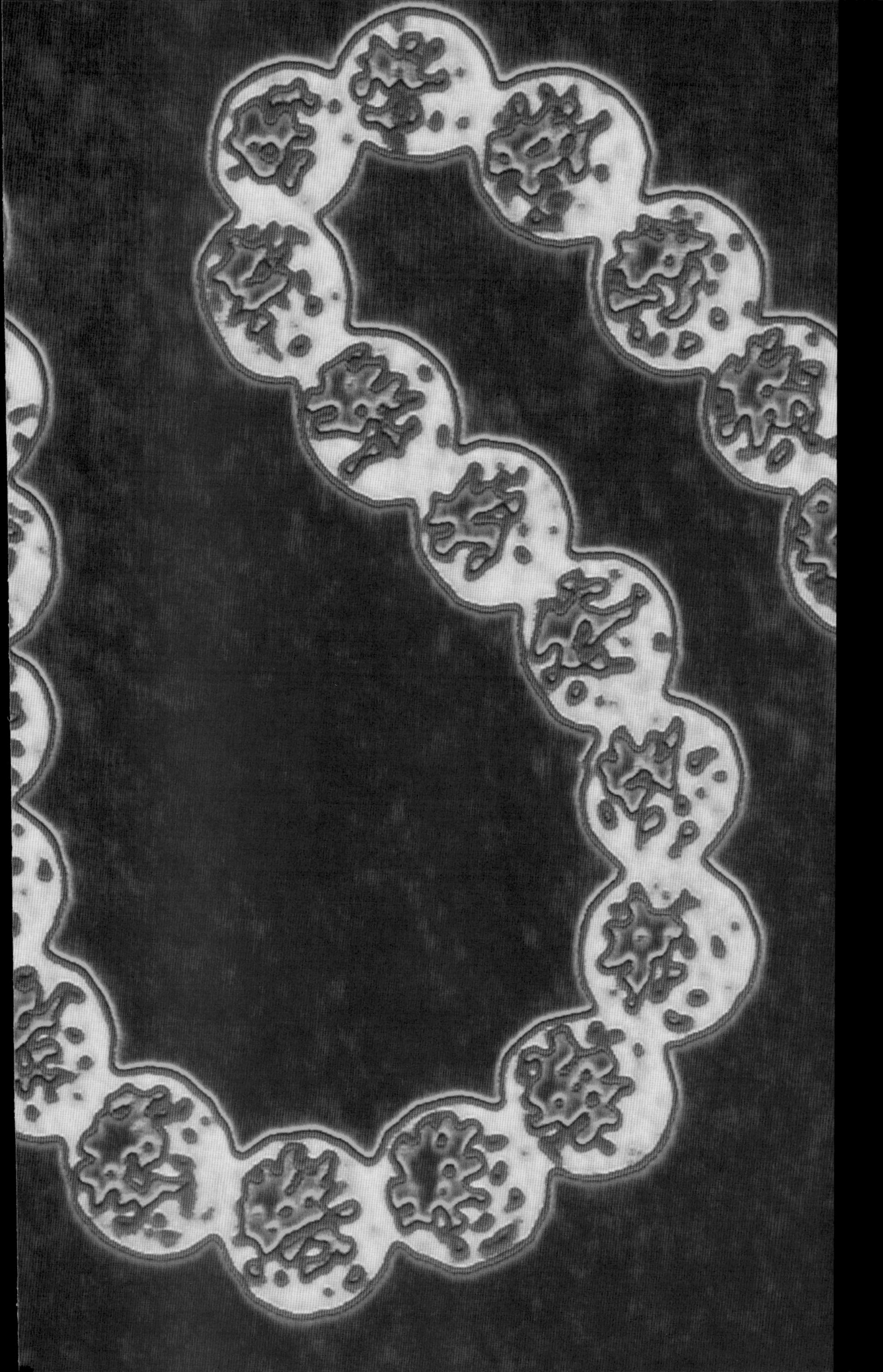

Left: *Streptococcus pyogenes* (coloured transmission electron micrograph)

The bacteria *Streptococcus pyogenes* is routinely found in the nose or throat of healthy humans. It multiplies to form chains – unlike staphylococcus bacteria which make grape-like clusters. Pathogenic (disease-spreading) strains of *Streptococcus pyogenes* cause infections of the skin (including impetigo); of the uterus following childbirth (puerperal sepsis); of the bloodstream (septicaemia); and scarlet fever. *Streptococcus pyogenes* accounts for most incidences of sore throats and tonsillitis in young children. The use of penicillin drugs is effective as a treatment. (Magnification: x7000 at 6x4.5cm/2¼x1¾in size)

Far left: Legionella bacteria (coloured transmission electron micrograph)

The bacteria *Legionella pneumophila* is the cause of Legionnaires' disease. Here, numerous bacteria are seen sectioned at different angles: typically, as at centre, they are rod-shaped. *Legionella pneumophila* was first identified as an agent of disease after a mysterious outbreak of pneumonia caused 29 deaths at an American Legion convention in 1976. The bacteria were found living in water-tanks, shower-heads and air-conditioning systems. They reproduce inside amoebae, where they are protected from attempts at sanitation such as chlorination. (Magnification: x20000 at 5x7cm/2x2¾in size)

Right: Streptococcus salivarius (coloured scanning electron micrograph)

These misshapen tennis balls are streptococcus bacteria. They colonize the mouth and upper respiratory tract within a few hours of our birth. *Streptococcus salivarius* is the first bacterium to settle on dental plaque and may accelerate tooth decay. It is otherwise normally harmless unless it enters the bloodstream, when it causes septicaemia. It resists the body's immune response by altering the proteins on its surface, effectively disguising itself and avoiding recognition. (Magnification: x5335 actual size)

Far right: E. coli bacteria (coloured scanning electron micrograph)

Looking more like cake decorations than bacteria in this image, rod-shaped *Escherichia coli* (commonly shortened to *E. coli*) are normal inhabitants of our intestines. They are usually quite harmless there and may even benefit us by producing vitamin K2 to combat other bacteria. Under certain conditions however *E. coli* multiplies so fast that our immune system cannot cope. Variations of it are responsible for gastroenteritis, particularly in tropical countries – 'traveller's diarrhoea' – and for 80% of all urinary tract infections. (Magnification: x3000 at 6x7cm/ 2¼x2¾in size)

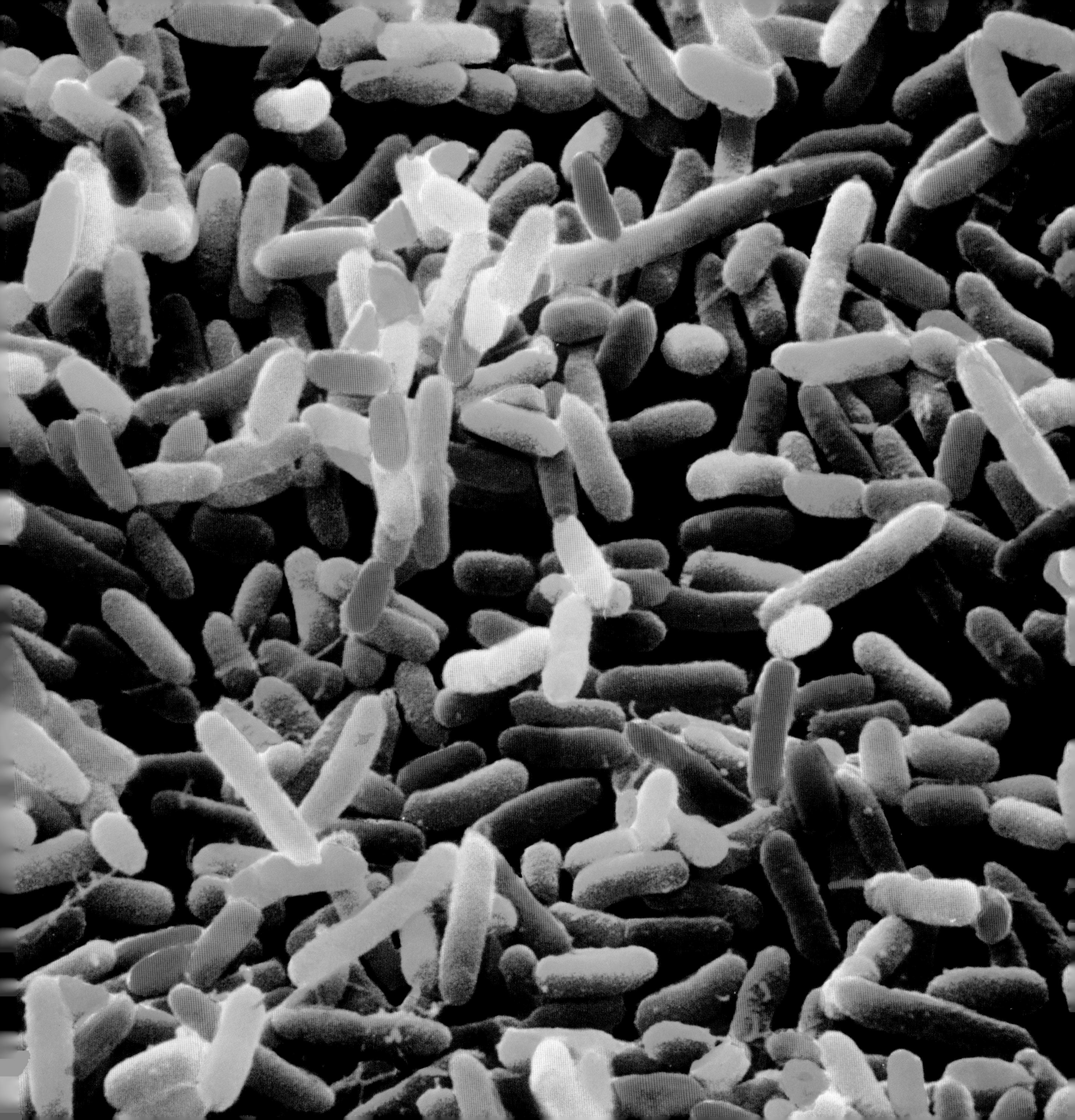

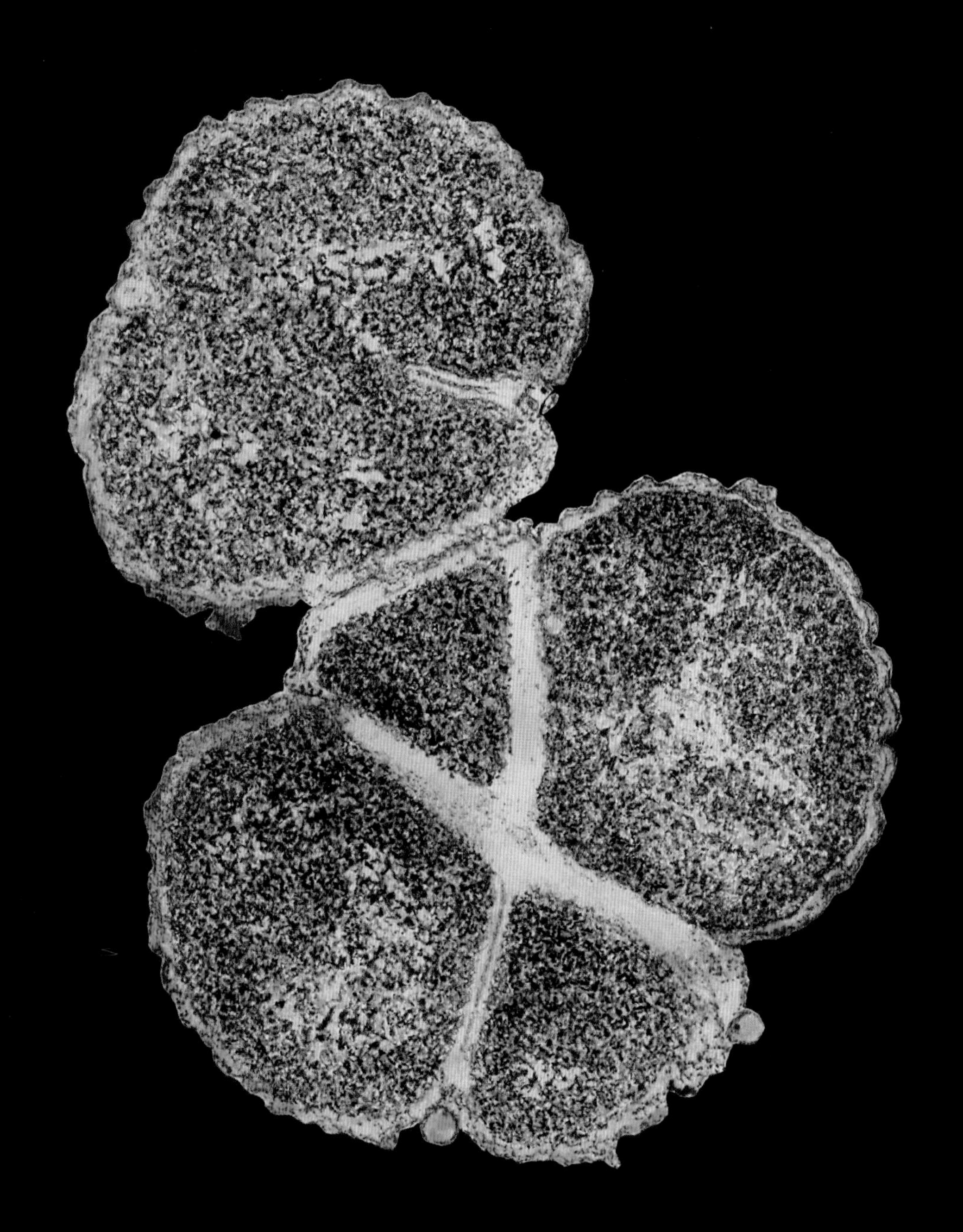

Above: ***Neisseria meningitidis***
(coloured transmission electron micrograph)
The bacteria *Neisseria meningitidis* causes meningococcal meningitis, an inflammation of the tissue which lines the brain and spinal cord. Symptoms include headache, fever, muscle rigidity and delirium. Treatment is with antibiotics. *Neisseria meningitidis* spreads through coughs and sneezes, and is an obligate human parasite – which means it depends on its human host to reproduce. In the image the lower cell has undergone a double cell division, producing four bacterial cells. The upper one is beginning to divide. (Magnification: x42500 at 6x7cm/2¼x2¾in size)

Right: Enlarged heart (coloured X-rays)
Two images of the heart (the light mass at the centre of the images), viewed from the right and left sides of the patient. The patient has taken a meal containing barium sulphate, which makes the throat (the bold orange ribbon) visible under X-ray. The X-rays also show that the left ventricle of the heart has become enlarged, like any muscle, because of an increased workload; here it is having to pump harder to overcome a weak mitral valve. (Magnification: unknown)

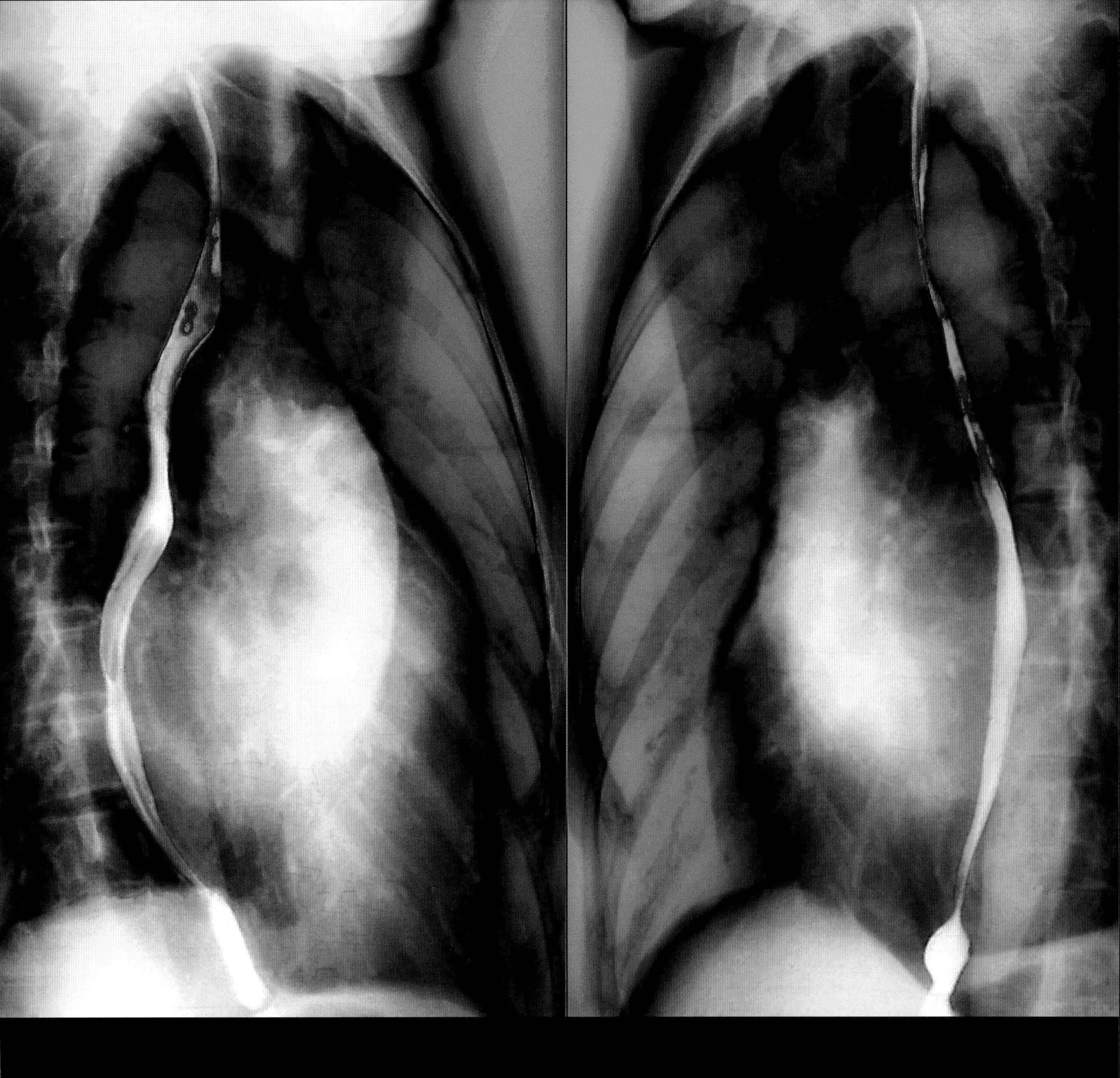

Human heart after a heart attack (coloured scintigram)
This image shows the uptake of a radio-isotope, Thallium-201, by the heart's muscle cells. Healthy heart muscle is coloured mauve, green and light blue. The obstruction of a coronary artery restricts blood flow, causing a heart attack and the death of a section of heart muscle – the dark red break or 'cold spot' between two blue areas at bottom left. This sort of scan is also used to monitor the performance of a heart at rest and during exercise. (Magnification: unknown)

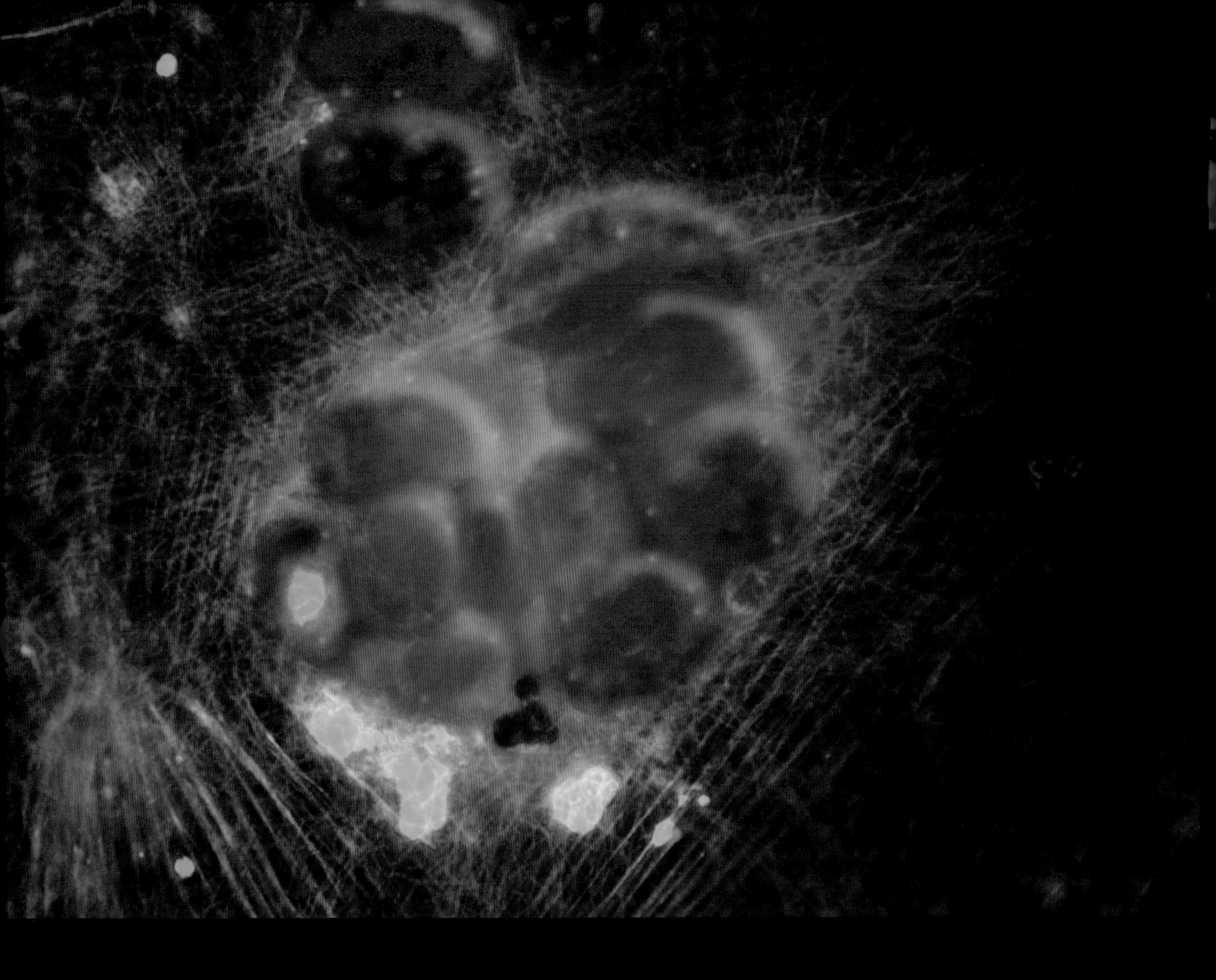

Herpes virus infected cell (immunofluorescence deconvolution micrograph)
Herpes viruses emerged some 200 million years ago. It's estimated that around 90% of all humans are infected with one of the 130 strains, which can remain in the body for life. There are three classes of herpes virus, and the image shows a cell infected by one of the gamma subgroup. The cell is multinucleate – it has several nuclei (blue) – and the cell body's structure of protein microtubules (red) and actin (green) is clearly visible (Magnification: unknown).

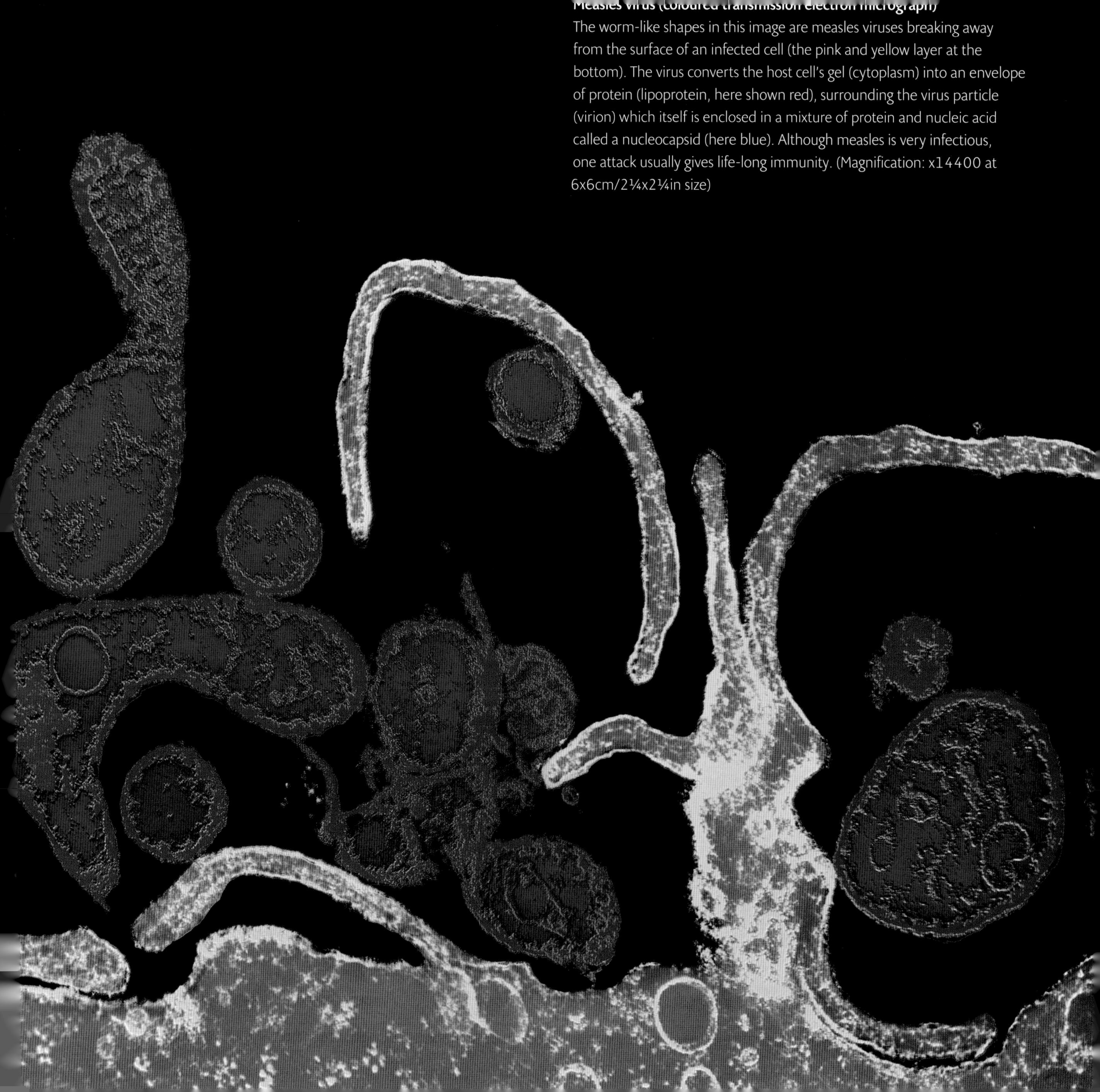

Measles virus (coloured transmission electron micrograph)

The worm-like shapes in this image are measles viruses breaking away from the surface of an infected cell (the pink and yellow layer at the bottom). The virus converts the host cell's gel (cytoplasm) into an envelope of protein (lipoprotein, here shown red), surrounding the virus particle (virion) which itself is enclosed in a mixture of protein and nucleic acid called a nucleocapsid (here blue). Although measles is very infectious, one attack usually gives life-long immunity. (Magnification: x14400 at 6x6cm/2¼x2¼in size)

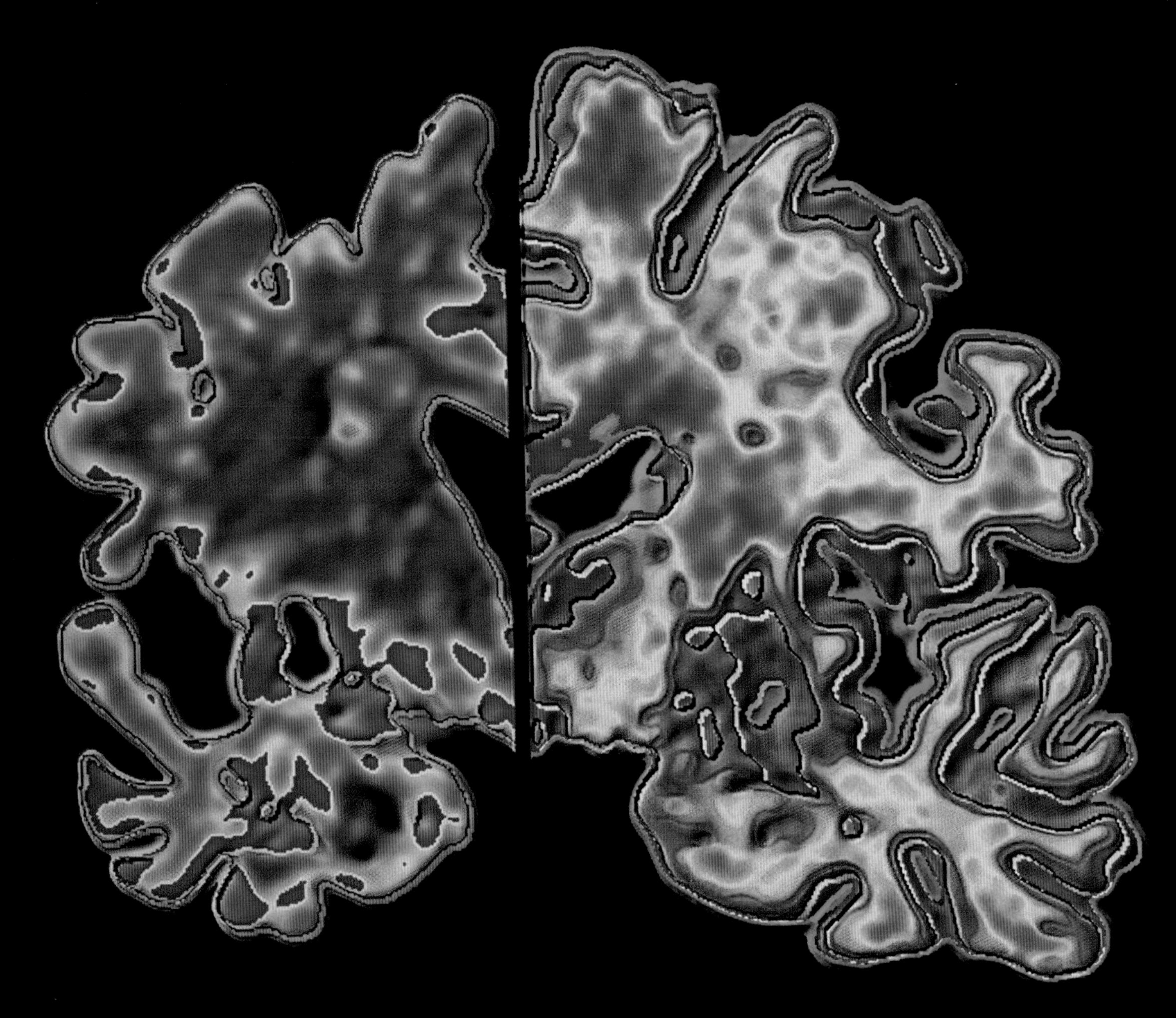

Above: Alzheimer's brain (computer-processed coronal magnetic resonance imaging scan)
Alzheimer's disease accounts for most cases of senile dementia. Symptoms include memory loss, disorientation, personality change and delusion. It ultimately leads to death. These vertical cross-sections compare the brain of an Alzheimer's patient (at left) with a healthy brain (at right). The Alzheimer's disease brain (brown) is considerably shrunken, due to the degeneration and death of nerve cells. Apart from a decrease in brain volume, the surface of the brain is often more deeply folded. (Magnification: unknown)

Right: Muscular dystrophy (confocal light micrograph)
This image shows a cross-section through muscle tissue affected by muscular dystrophy, a genetic disorder which causes muscle-wasting and loss of function. The tissues in the image include: muscle fibres (purple), mesenchymal tissue (green and yellow) which supports muscle fibre, and fat (adipose) tissue (black). This example illustrates adipose metaplasia, where muscle is being replaced by fat. This is a characteristic symptom of muscular dystrophy. (Magnification: unknown)

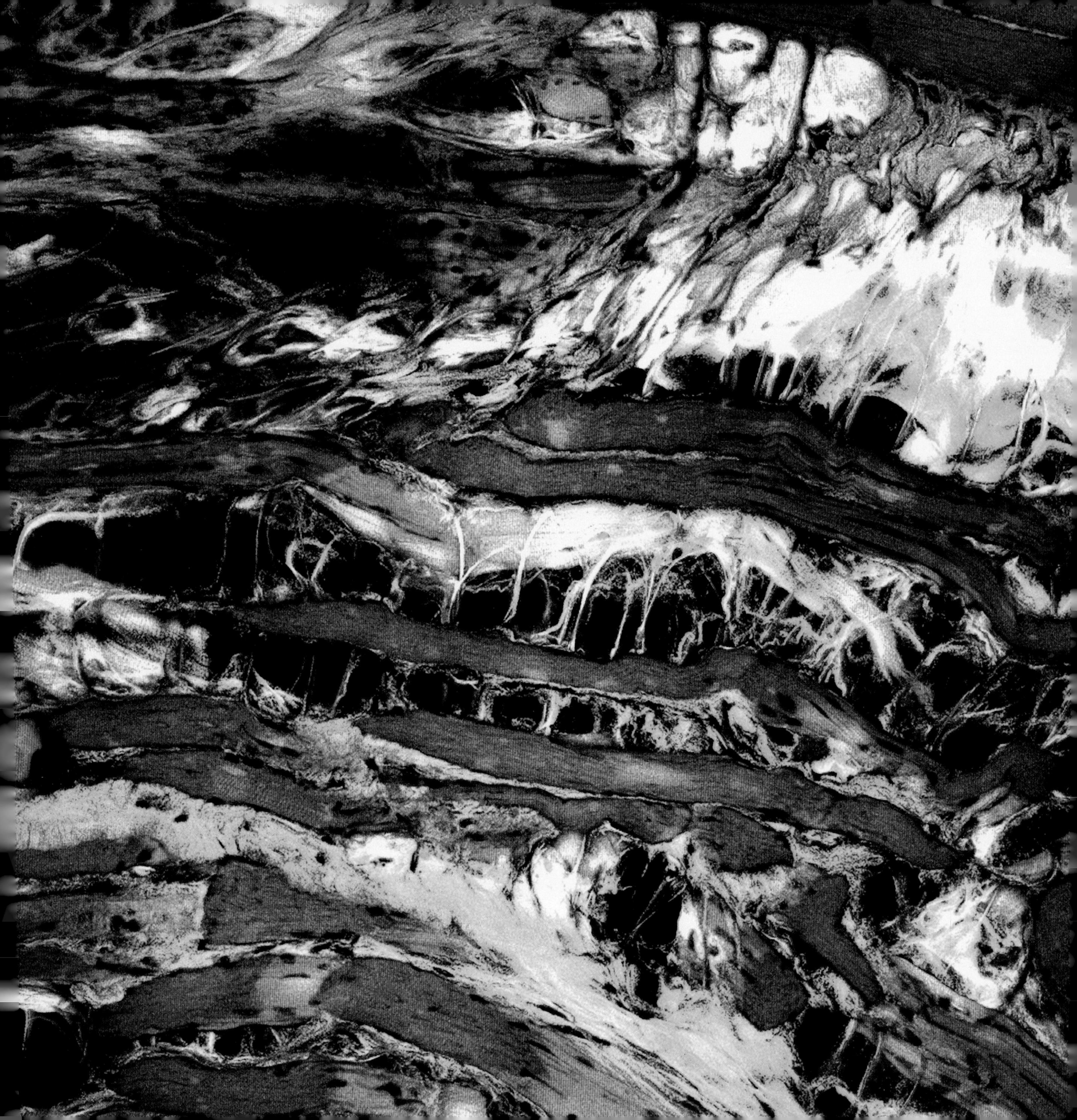

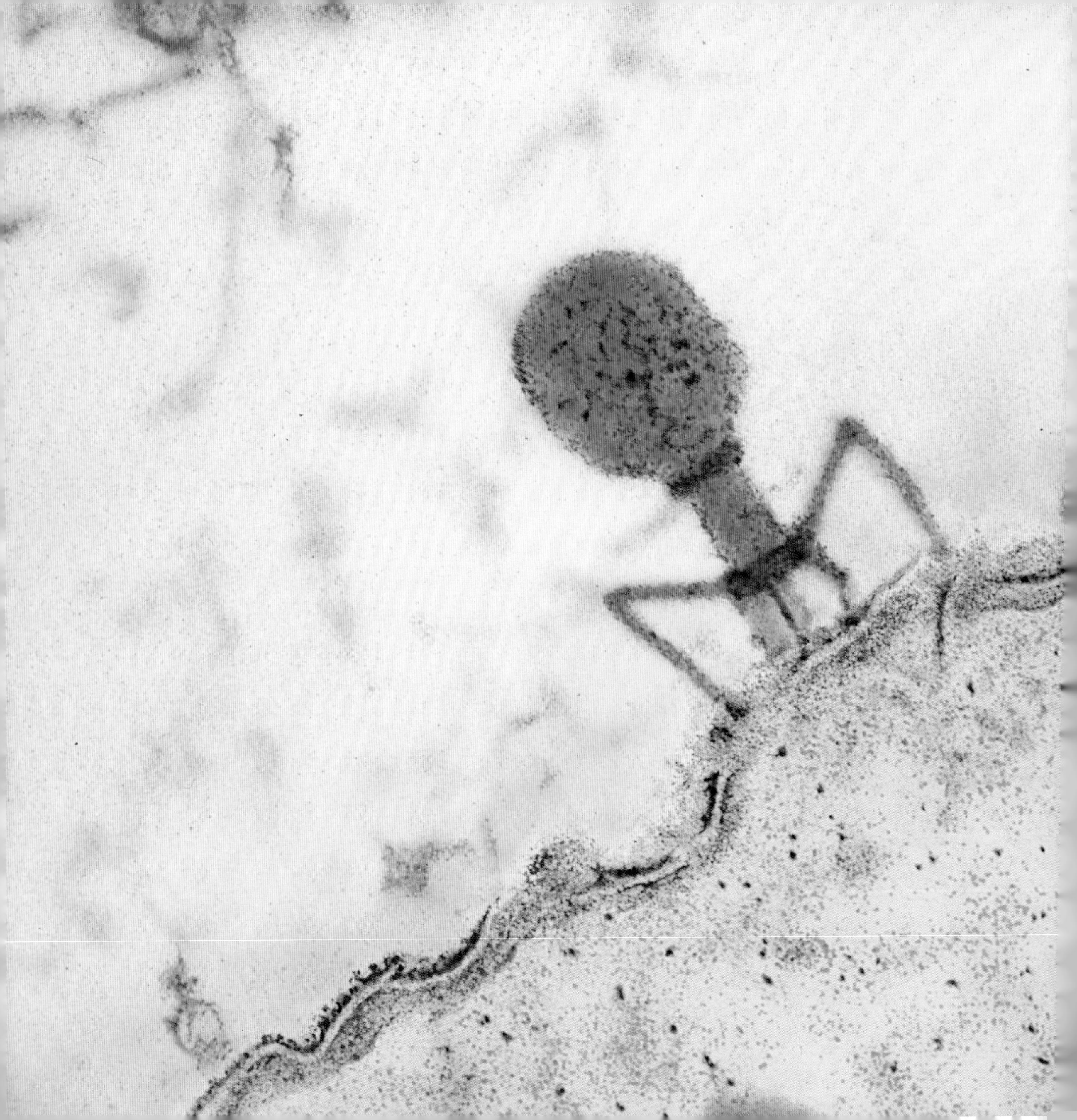

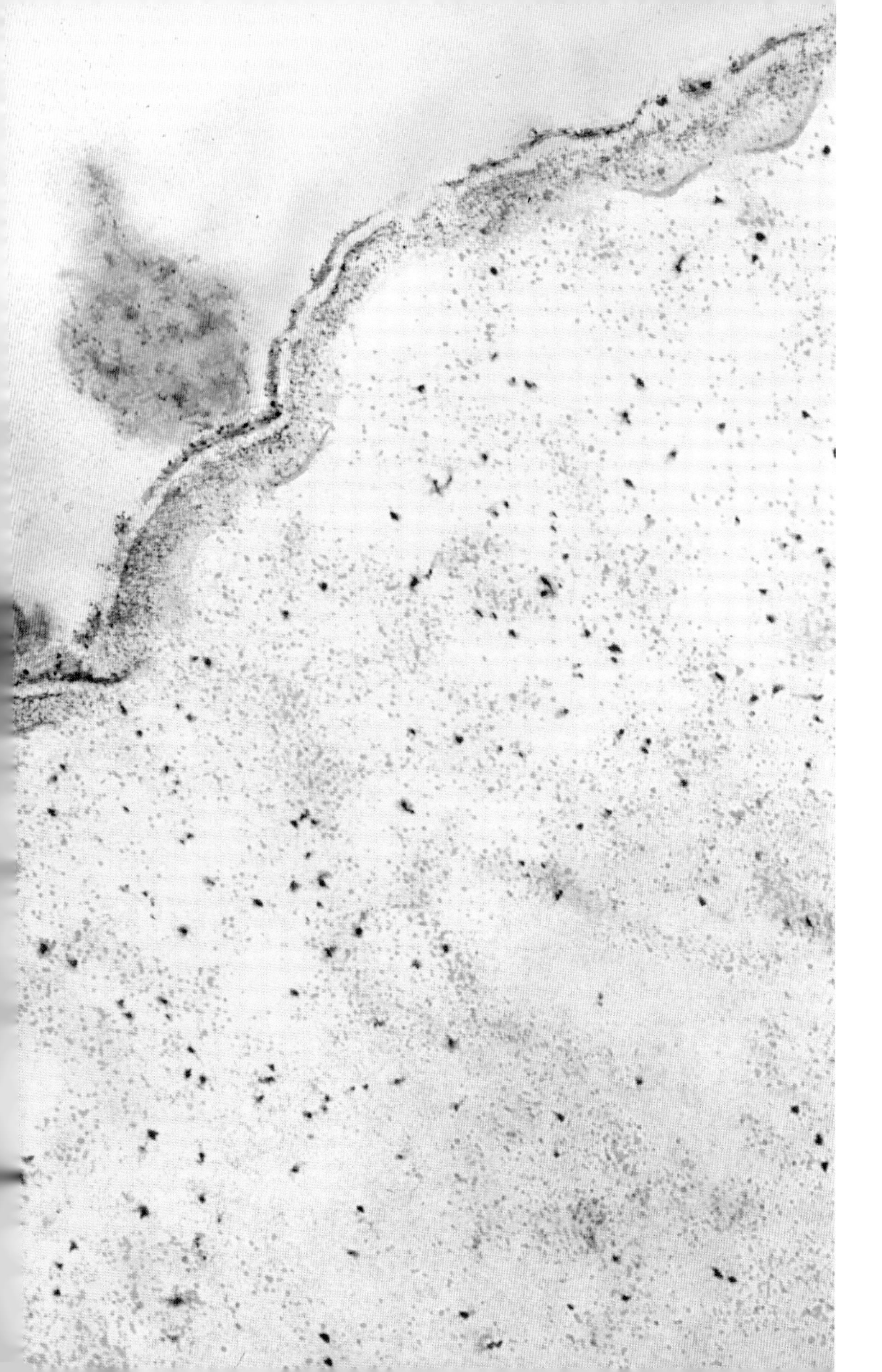

Bacteriophage (coloured transmission electron micrograph)

Bacteriophages are viruses that infect bacteria; this one, a T4 bacteriophage (orange), has just injected its viral DNA into an *E. coli* bacterium (blue). It is anchored to the surface of the cell by spidery tail fibres. The tail contracts to allow a syringe-like tube below its base to puncture the cell membrane, emptying the DNA contents of the head into the bacterium. New T4 phages then grow, kill and depart from the host cell within 30 minutes. (Magnification: x65000 at 6x7cm/2¼x2¾in size)

Right: Giardia protozoan (coloured scanning electron micrograph)
A parasite of the intestinal tract, the giardia protozoan (purple) is most common in tropical regions, where it causes the disease giardiasis. It is spread through contaminated food and water and causes abdominal pain, diarrhoea and nausea. Giardia protozoans inhabit the gullet and flatten its lining (green), affecting digestion. It may be able to pass between humans and other species including dogs, cats, cattle, horses – and beavers, hence its colloquial name in some areas: beaver fever. (Magnification: x6000 at 10cm/4in size)

Far right: Salmonella bacteria (coloured scanning electron micrograph)
These rod-shaped bacteria (pink) are an army of *Salmonella typhimurium* bacteria, growing on blood agar in the laboratory. Agar is a medium on which bacteria are nurtured to study their behaviour. Their thread-like flagellae (green) are used for movement. *Salmonella typhimurium*, transmitted in food, is a major cause of food poisoning (salmonellosis) in humans. Infections are common in poultry and eggs. Inadequately cooked pork and beef are also potential sources, and rats and mice are common carriers. (Magnification: x3900 at 10cm/4in size)

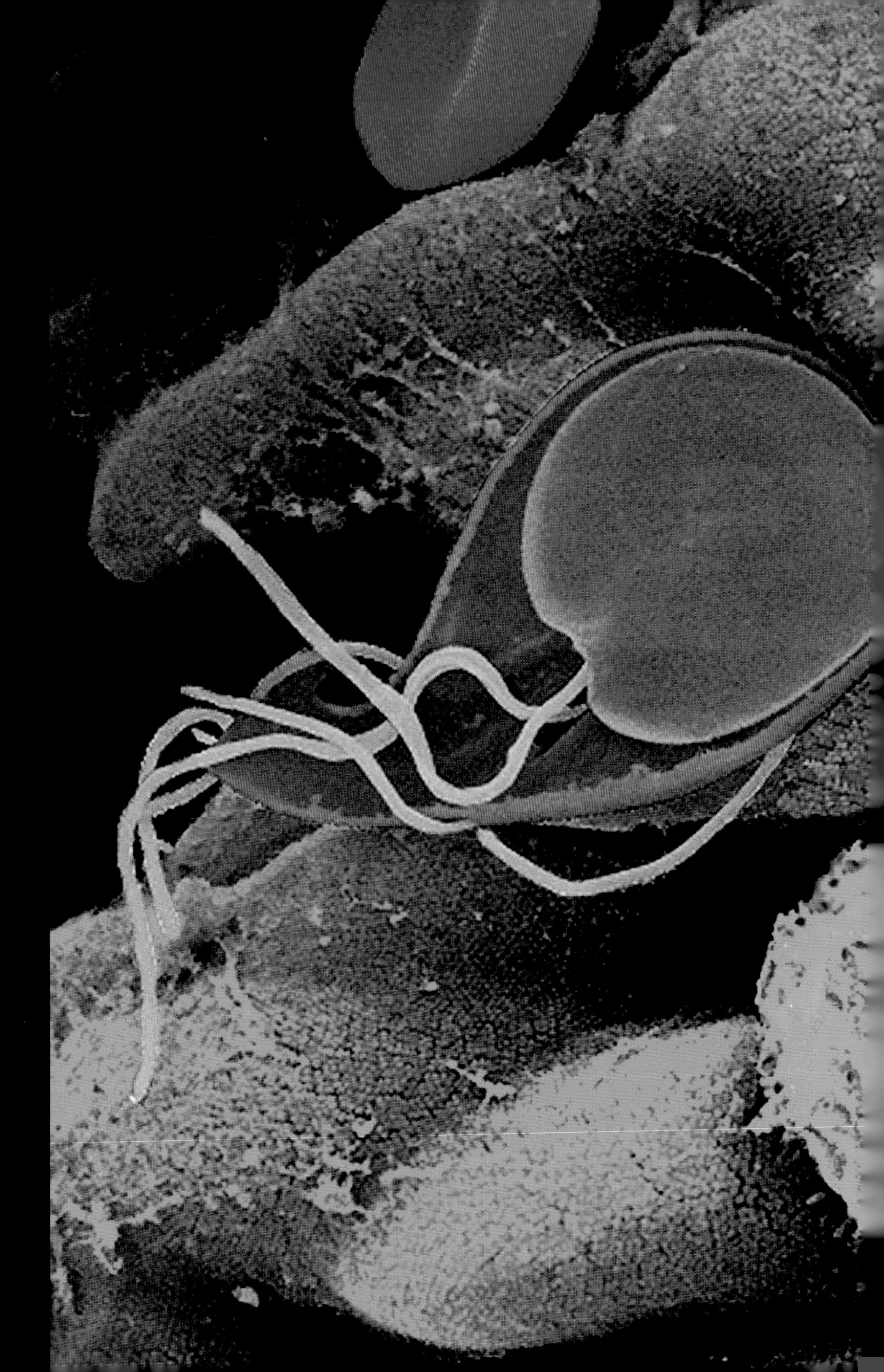

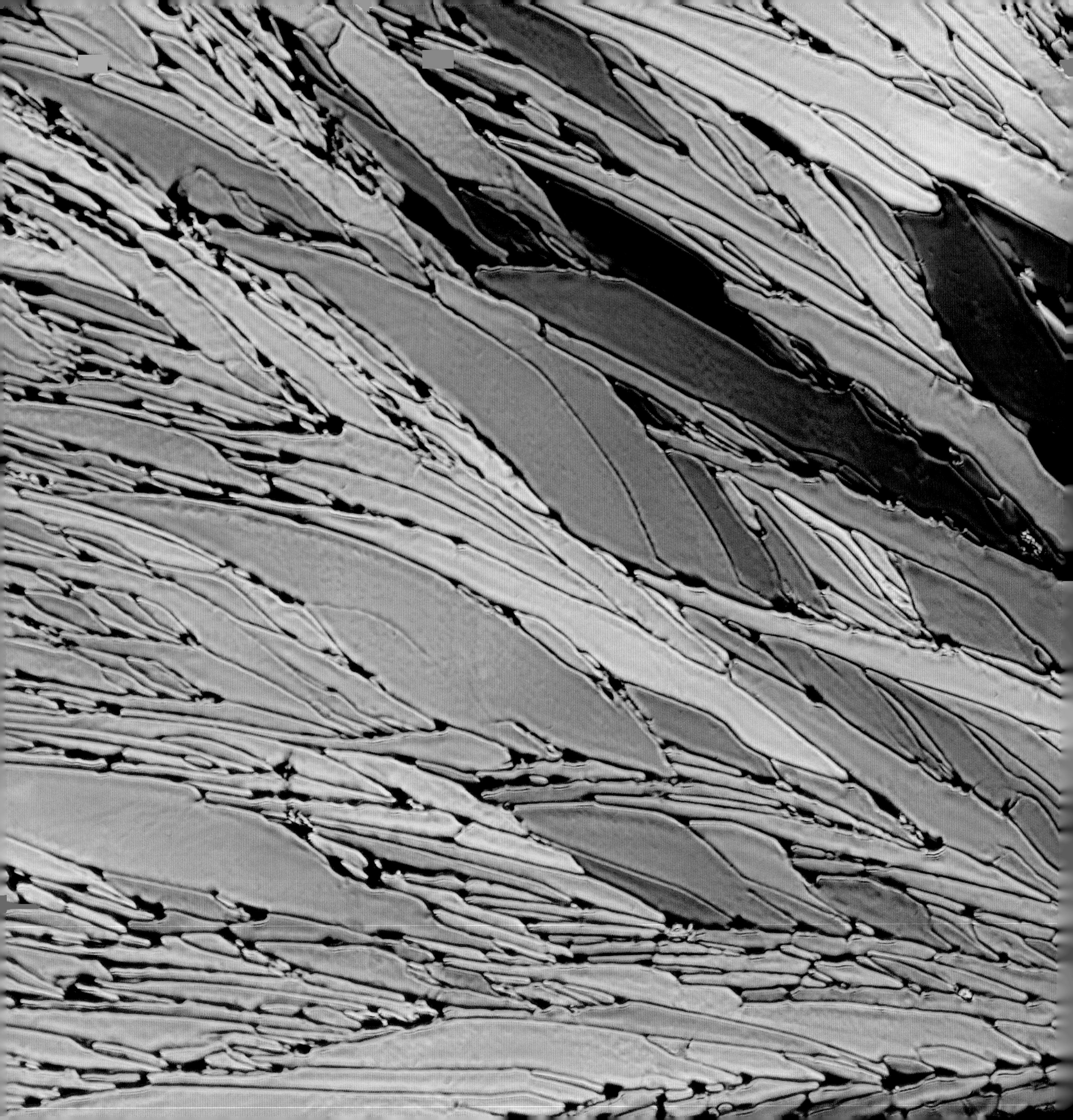

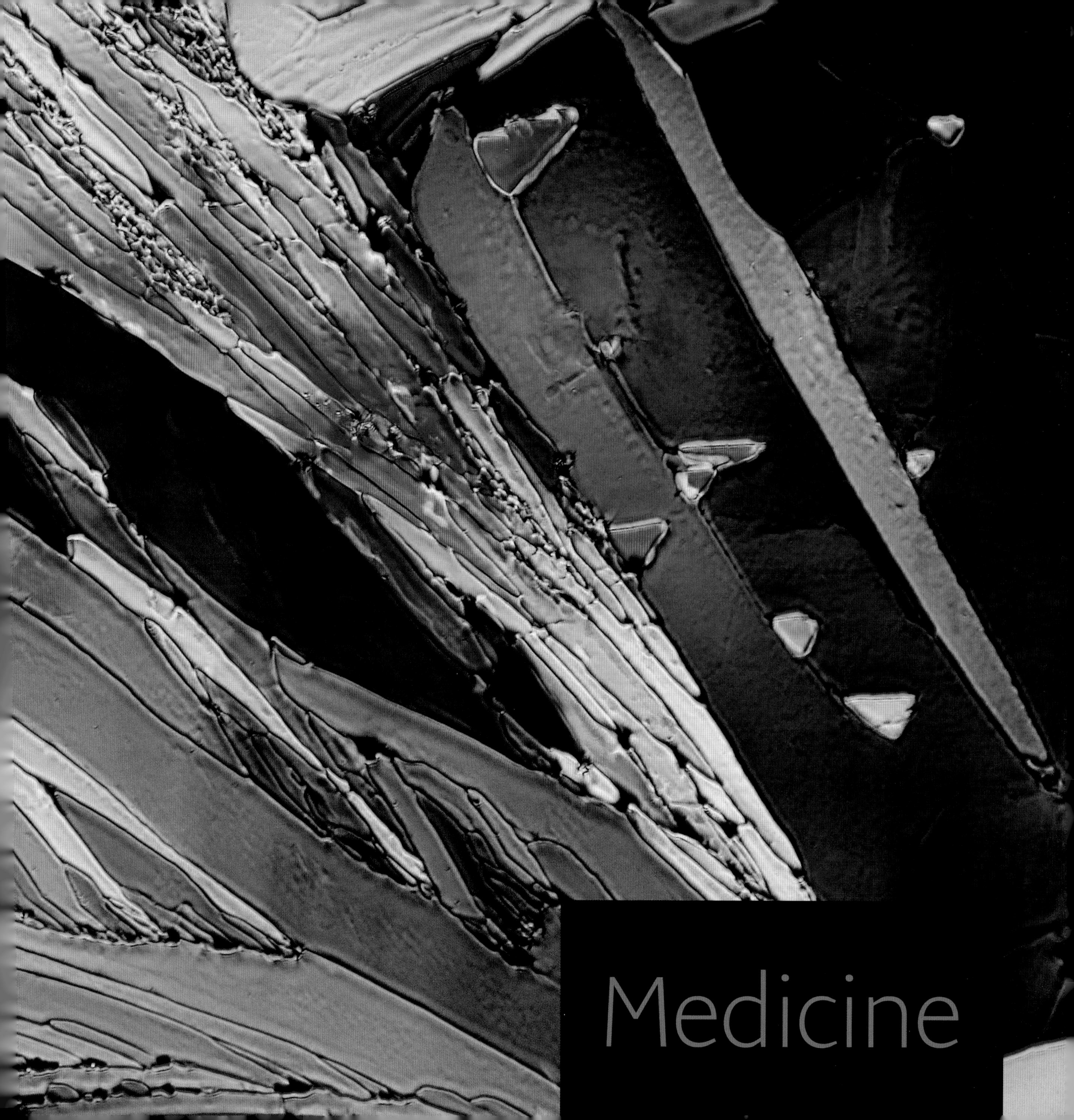

Medicine

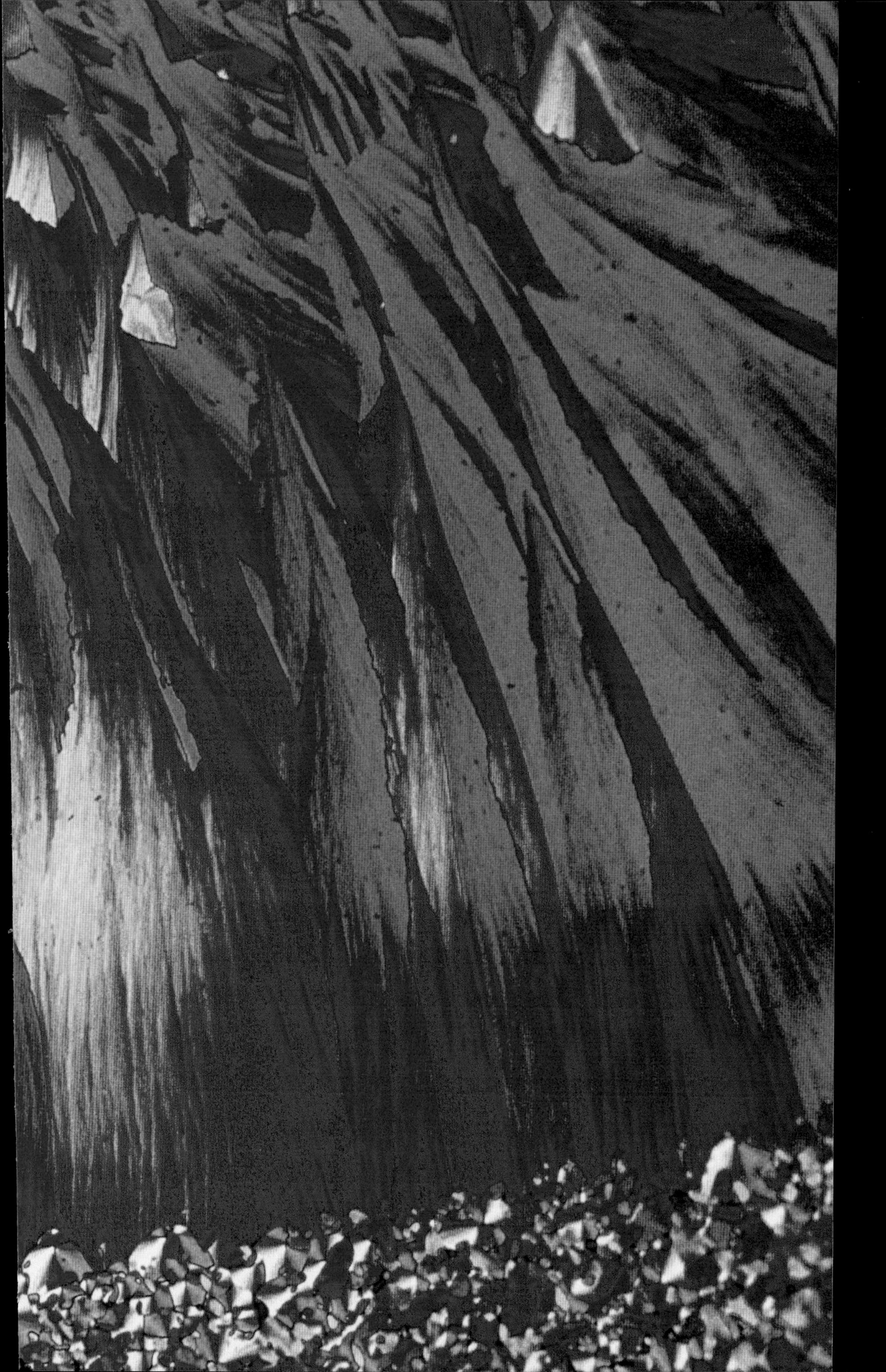

Previous page: Metformin crystals (polarized light micrograph)
These curving shards of stained glass are crystals of the diabetes drug metformin. Metformin is used to treat type 2 diabetes (also known as non-insulin-dependent diabetes mellitus). It controls blood sugar levels by reducing the production and secretion of glucose by the liver. It also helps reduce the risk of heart disease in obese diabetes patients, by lowering levels of low-density lipoprotein (so-called 'bad cholesterol') and triglycerides in the blood (which carry saturated fat). (Magnification: x220 at 10cm/4in size)

Left: Taxol crystals (polarized light micrograph)
First isolated in 1964, taxol is a drug used in chemotherapy. It is proving effective against some cancerous tumours, notably in the lungs, breasts and ovaries. Taxol is a naturally occurring substance: it is produced by fungus within the bark of the Pacific Yew (*Taxus brevifolia*). The tree is only found on the Pacific coast of North America, where it is already an endangered species. Research efforts are therefore being focussed on synthesizing the drug in the laboratory. (Magnification: x25 at 6x7cm/2¼x2¾in size)

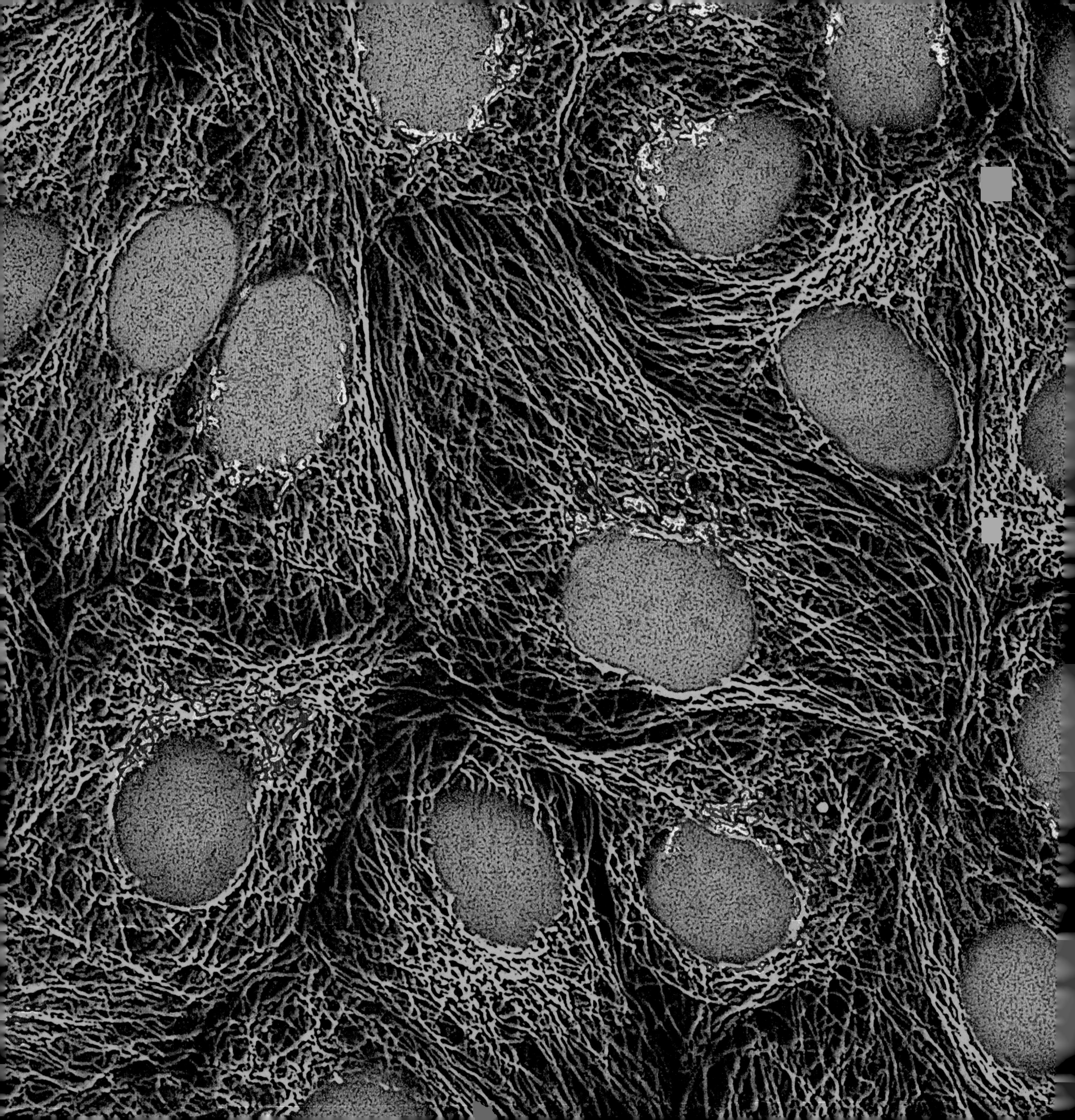

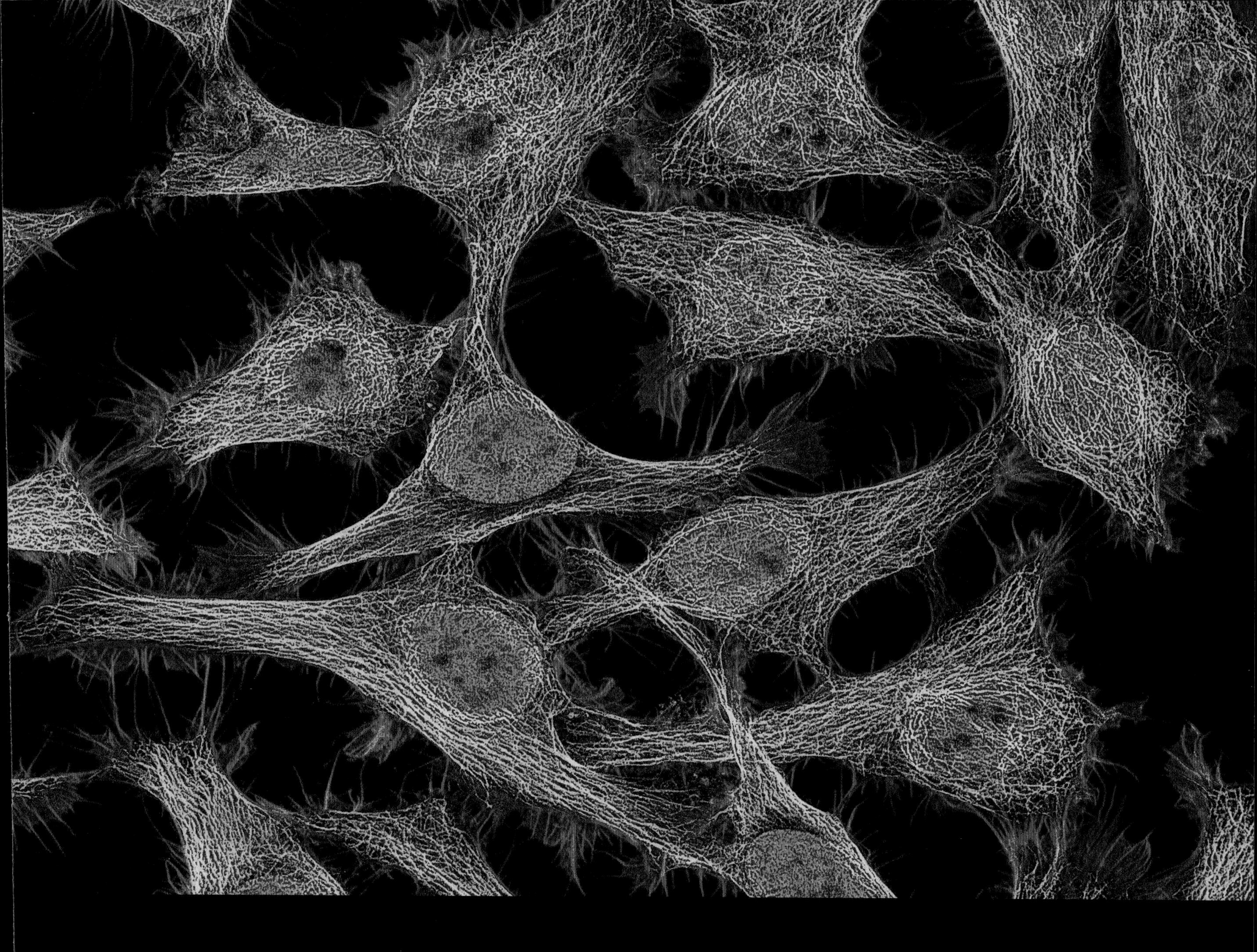

Left: HeLa cells (multiphoton fluorescence light micrograph)
HeLa cells are human cancer cells grown in the laboratory for medical research. Under the right conditions they can reproduce themselves without limit, and are considered immortal. In this image the cell nuclei, which contain the cells' genetic information, are blue. Golgi bodies, which modify and package proteins, are orange. Microtubules, protein filaments that make up part of the cytoskeleton, are green. The cytoskeleton maintains the cells' shape, allows some cellular mobility and is involved in intracellular transport. (Magnification: unknown)

Above: HeLa cells (multiphoton fluorescence light micrograph)
HeLa cancer cells are an important research tool. They are propagated from cells originally removed from Henrietta Lack, who suffered from cervical cancer and died in 1951. They are immortal: they can divide and multiply without limit in a sustaining laboratory environment. They therefore form a stable basis for continuing research. In this image the cells' nuclei are purple. Microtubules (blue) and actin microfilaments (red) are protein filaments that make up part of the cells' structural skeleton. (Magnification: unknown)

Right: Migraleve crystals (polarized light micrograph)
Migraine sufferers are familiar with the drug Migraleve. The treatment consists of two tablets, a pink one (taken first) and a yellow one. The crystals in this image are in the pink tablet, which consists mostly of paracetamol (96%) with small amounts of codeine phosphate, buclizine hydrochloride and dioctyl sodium sulphosuccinate. The pink pill addresses the nausea of migraine, while the yellow one confronts the headache. (Magnification: x8 at 3.5cm/1¼in size)

Far right: Aspirin crystals (polarized light micrograph)
Otherwise known as acetyl-salicylic acid, Aspirin is an analgesic (literally 'anti-pain') drug used for the relief of moderate pain, the reduction of inflammation and of fever. Aspirin is derived from salicylic acid, the active ingredient of willow bark, which in the mid-eighteenth century was chewed to relieve symptoms of malaria. Salicylic acid is effective but highly unpalatable, so chemists developed acetyl-salicylic acid. Aspirin helps prevent blood clotting and thus may play a role in reducing coronary heart disease. (Magnification: x10 at 3.5cm/1¼in size)

Left: Aspirin crystals (polarized light micrograph)
The active ingredient of aspirin is salicylic acid, derived from the bark of the willow tree (whose botanical genus is *Salix*). The curative properties of the tree were known in ancient Egypt, and the acid was first isolated in 1763 by English vicar Edward Stone. Synthetic aspirin was first produced by Felix Hoffmann of the German pharmaceutical company Bayer in 1897. It is widely used for minor fevers, aches and pains – 40,000 tonnes (44,093 tons) of it are consumed annually worldwide. (Magnification: unknown)

Above: Allantoin crystals (polarized light micrograph)
Human beings pass uric acid when they urinate, unlike most other vertebrates who convert the uric acid to allantoin. However, allantoin is present in foetal urine and in human amniotic fluid (which protects the unborn foetus). Its presence there promotes rapid cell regeneration; and allantoin is therefore used as an ingredient in creams for healing wounds and burns. It also has a skin-softening effect and is used in many types of cosmetics. (Magnification: unknown)

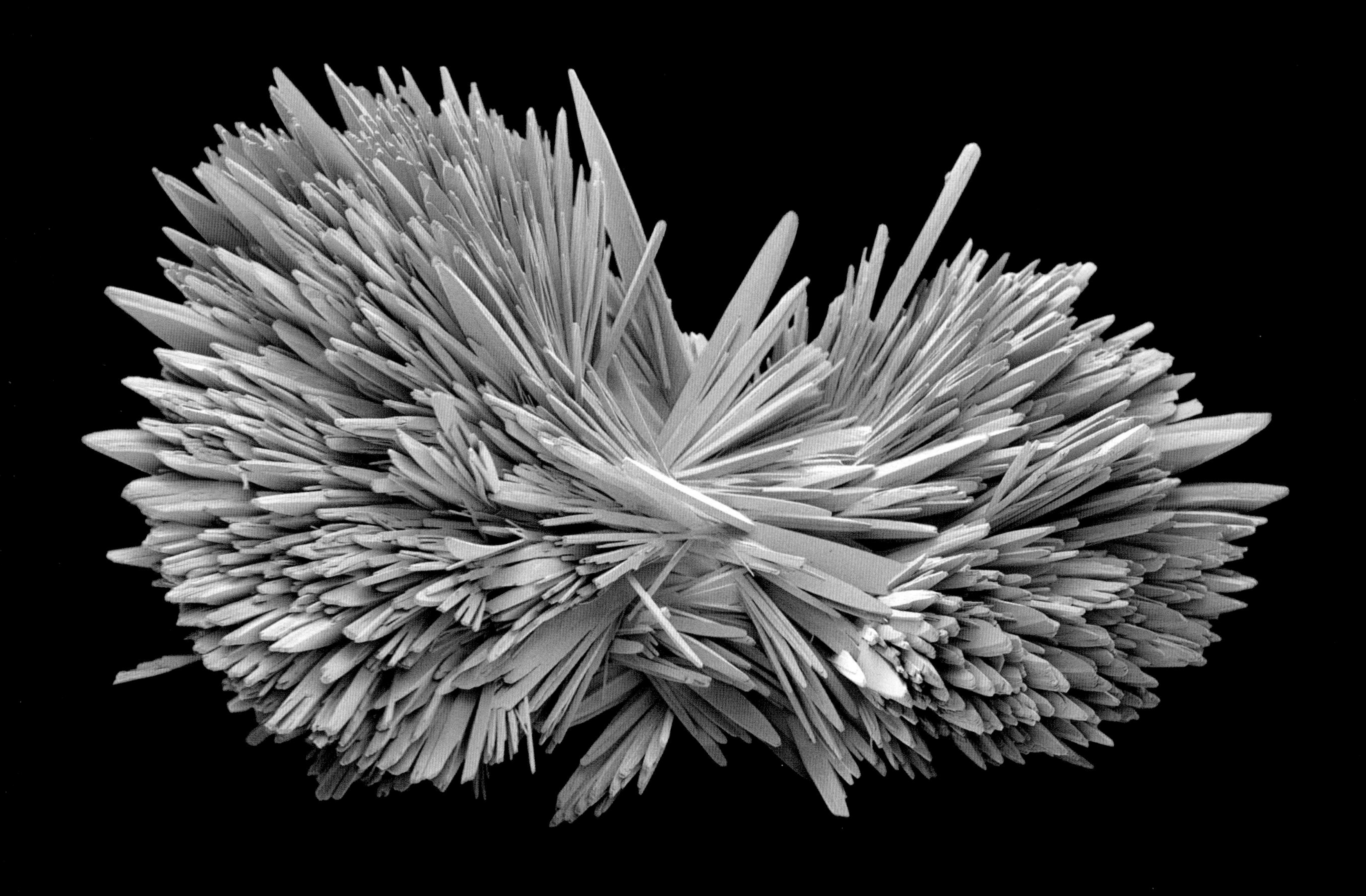

Left and above: Ventolin crystals (coloured scanning electron micrographs)
Asthma is a condition that causes narrowing of tiny airways in the lungs, which become inflamed and lead to attacks of wheezing and breathlessness. These are crystals of salbutamol sulphate, which is marketed under several brand names including Ventolin as a drug used to treat asthma. Salbutamol sulphate works by mimicking adrenaline to stimulate muscles around constricted airways in the lungs. The muscles then relax, opening the airways and relieving asthma attacks. Although mostly used by asthma sufferers, salbutamol sulphate also treats other conditions. For example, premature labour can be delayed by a similar relaxation of the muscles of the uterus. It has also provided relief from cystic fibrosis and hyperkalemia. (Magnification unknown)

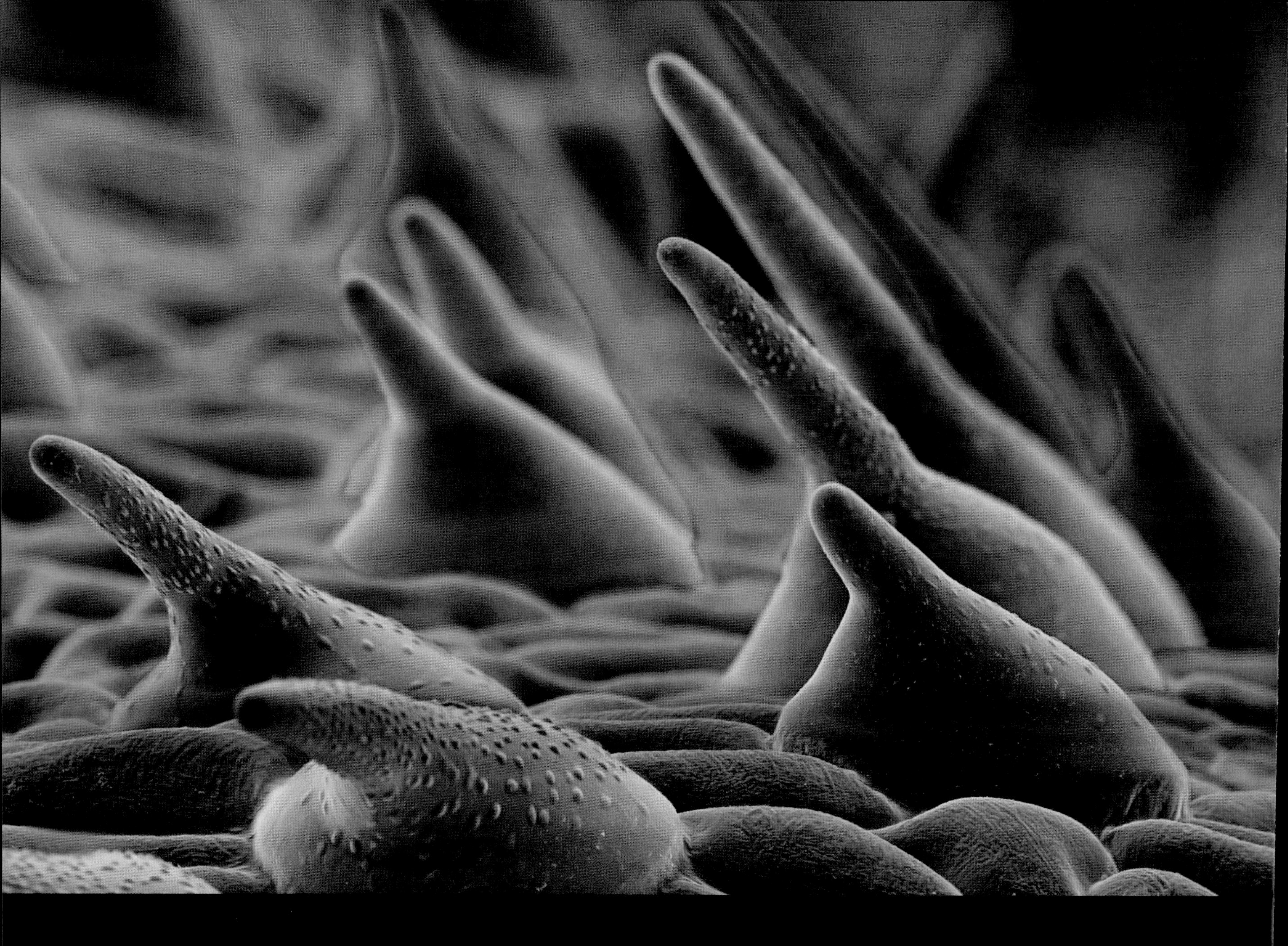

Above: Cannabis leaf trichomes (coloured environmental scanning electron micrograph)

In botany, trichomes are the hairs on the stems and leaves of plants. These trichomes on the surface of a cannabis (*Cannabis sativa*) leaf are glandular – that is, they secrete a resin containing tetrahydrocannabinol, the active component of cannabis when used as a drug. Short-term medical use of cannabis can alleviate pain for those suffering chronic conditions such as fibromyalgia and rheumatoid arthritis, and nausea in patients with AIDS and those undergoing chemotherapy. Long-term use may affect memory and cognition. (Magnification: unknown)

Right: *Penicillium* spores (coloured scanning electron micrograph)

This chain of spherical spores, called conidia, is the result of reproduction by the *Penicillium* fungus. The spores are asexual – they are produced by one parent fungus without the need of another. *Penicillium* is a large group of fungi with over 300 species, most of which feed on decaying matter. Some species produce the penicillin antibiotics, and were the original source of these drugs. Others, for example *Penicillium camemberti* and *Penicillium roqueforti*, are used in the making of cheese. (Magnification: unknown)

***Penicillium* fungus (coloured scanning electron micrograph)**

Like a tray of flowers in a florist's shop, this image displays stalks of the fungus *Penicillium*. Specialized threads (hyphae, pink), called conidiophores, end in bunches of spores (conidia, yellow), the fungal reproductive units. The antibiotic penicillin is obtained from certain types of *Penicillium* fungi. It was discovered accidentally by Alexander Fleming in 1928. Its effectiveness was proved in the treatment of infected wounds in World War II, and won him a Nobel Prize in 1945. (Magnification: x75 at 10cm/4in size)

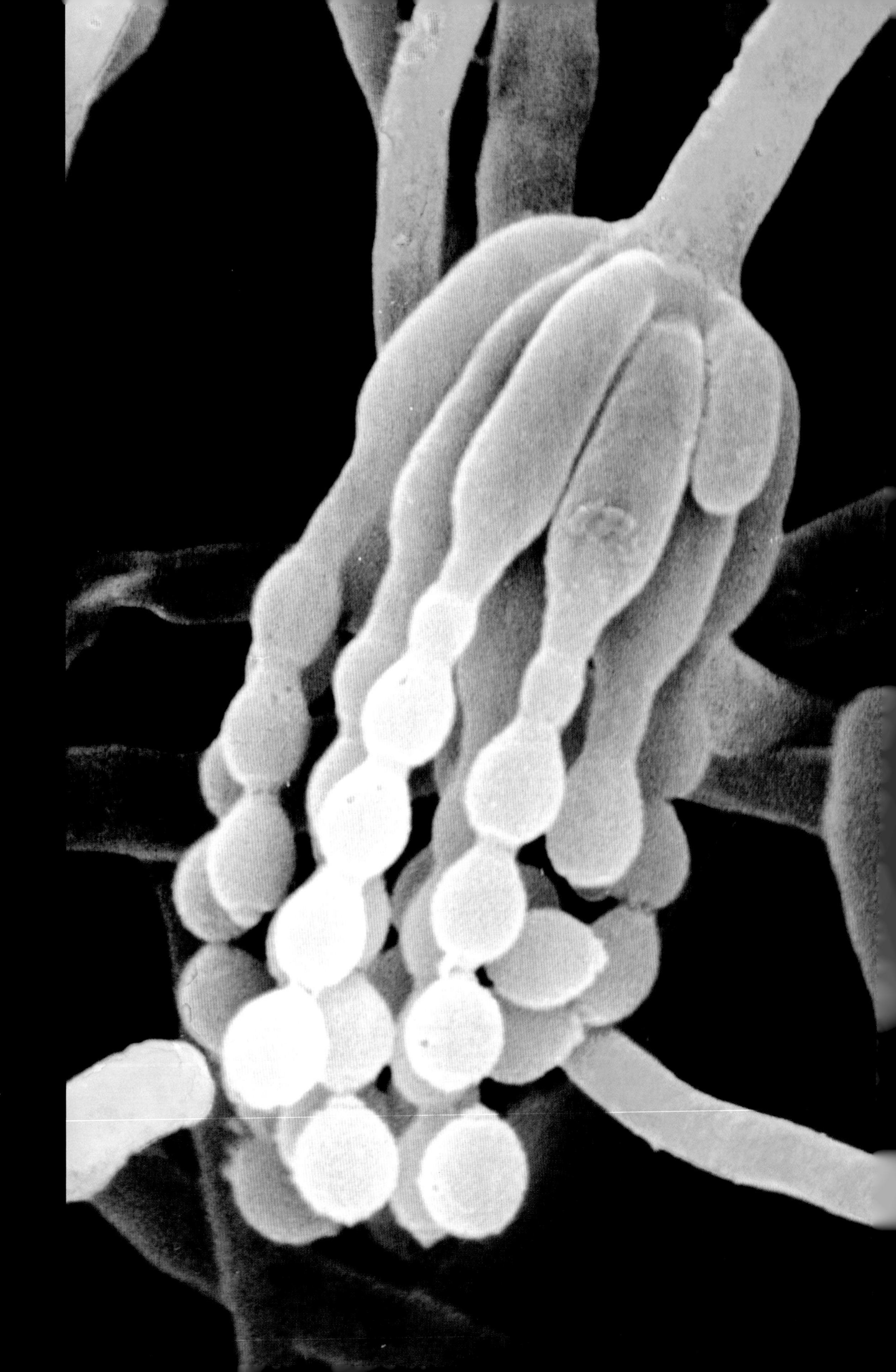

***Penicillium* (coloured scanning electron micrograph)**
The fungus family which gives us penicillin is the same one which grows as mould on bread. In this image the strands (hyphae) of *Penicillium* have been tinted green and the chains of spores (conidia) are blue. As the spherical spores ripen they detach from the parent fungus to germinate as new organisms. Penicillin was first produced from *Penicillium notatum*, effective against a range of bacteria; modern antibiotics are now more selective, targeted against specific micro-organisms. (Magnification: unknown)

***Penicillium* (coloured scanning electron micrograph)**

Penicillium is a fungus which produces the antibiotic drug penicillin. The drug was discovered accidentally by Alexander Fleming in 1928 when his bacterial culture dish became contaminated with the fungus. The discovery revolutionized medicine and won Fleming the Nobel Prize in 1945. The fruits of the fungus grow as chains of round spores (conidia, the pink 'flowers' in the picture) on fine stems branching from the main thread (the hypha). Each spore can germinate into a new fungus. (Magnification: unknown)

***Penicillium* fungus (differential interference contrast light micrograph)**
This fungus grew on a lemon. The round structures seen here are conidia, the asexual reproductive spores of *Penicillium*. Each chain of conidia grows on a specialized hypha (stalk) of the mature fungus known as a conidiophore. Many *Penicillium* species play a role in antibiotics and biotechnology. Others have a function in food production, including cheese and – in the case of *Penicillium nalgiovense* – to enhance the flavour and preservation of cooked meats. (Magnification: unknown)

Paracetamol crystals (polarized light micrograph)
These radial rosettes have been crystallized from a solution of paracetamol (also known as acetaminophen) in water. Paracetamol is an analgesic (painkiller) and antipyretic (reduces fever). The name derives from its full chemical name, PARa-ACETyl-AMino-phenOL. It is a metabolite (a chemical alteration during digestion) of a drug with similar effects called phenacetin. Phenacetin was banned in 1983 when it was found to raise the risk of certain cancers. Paracetamol has no such risks in therapeutic doses. (Magnification: unknown)

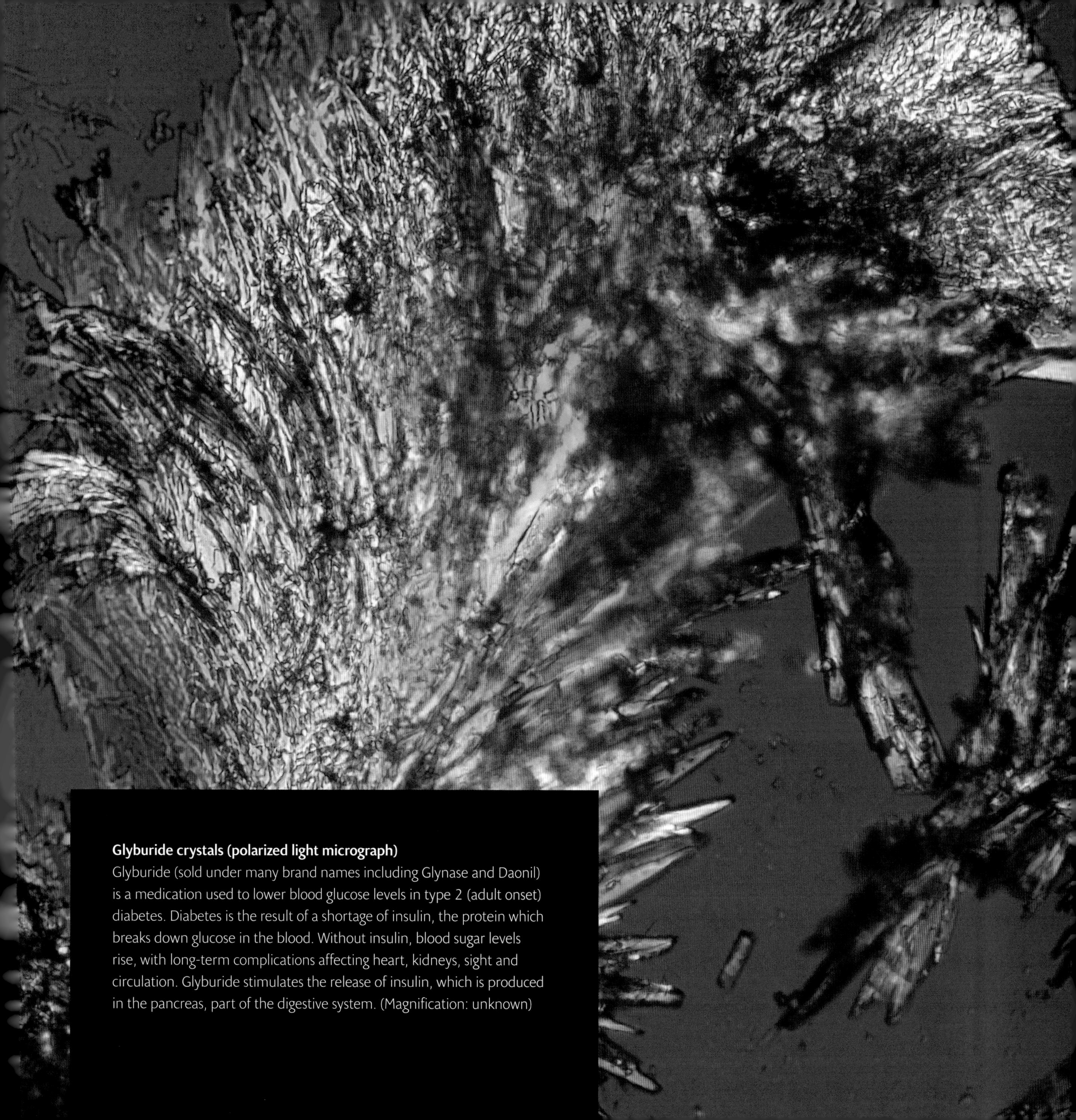

Glyburide crystals (polarized light micrograph)

Glyburide (sold under many brand names including Glynase and Daonil) is a medication used to lower blood glucose levels in type 2 (adult onset) diabetes. Diabetes is the result of a shortage of insulin, the protein which breaks down glucose in the blood. Without insulin, blood sugar levels rise, with long-term complications affecting heart, kidneys, sight and circulation. Glyburide stimulates the release of insulin, which is produced in the pancreas, part of the digestive system. (Magnification: unknown)

Didanosine crystal (light micrograph)
When HIV infects a cell, an enzyme adapts the virus's genetic template to make viral DNA, in a process called reverse transcription. The viral DNA is then integrated into the infected human's DNA, where normal cellular processes reproduce the virus. Didanosine acts to block reverse transcription, thus preventing HIV from multiplying. In 1991 it was only the second drug to be approved for HIV treatment, four years after the first, azidothymidine (AZT). (Magnification: unknown)

Erythromycin crystal (light micrograph)
Like penicillin, erythromycin is an antibiotic derived from a bacterium. It is frequently prescribed for those who are allergic to penicillin, and in some cases such as infections of the windpipe it is more effective. Erythromycin is produced from the *Saccharopolyspora erythraea* bacterium. It has a complex chemical structure which makes synthetic production very difficult. In the body it is quickly absorbed by white blood cells which carry it to the site of infection. (Magnification: unknown)

Right: Drug delivery microspheres (coloured scanning electron micrograph)
These microscopic spheres (green) are used to deliver drugs (pink) to specific sites in the body. The microspheres, made from a polymeric substance, are biodegradable, and break down to release the drugs they are carrying. They are usually made of polystyrene, which absorbs proteins easily and permanently. By applying the right protein, drug manufacturers can convince our immune systems to direct the microspheres and the drug they carry to the part of the body where they are needed. (Magnification: unknown)

Far right: Graft in artery (coloured computed tomography scan)
This graft has repaired the patient's aorta (the largest artery in the body) after the discovery of a dangerous aneurysm (a blood-filled swelling) in the aorta wall. If the swelling bursts it can quickly be fatal. Treatment involves removing the weakened part of the artery and replacing it with an artificial graft, which is held open by a splint-like device called a stent. CT scans like this one use X-rays to construct images of parts of the body. (Magnification: unknown)

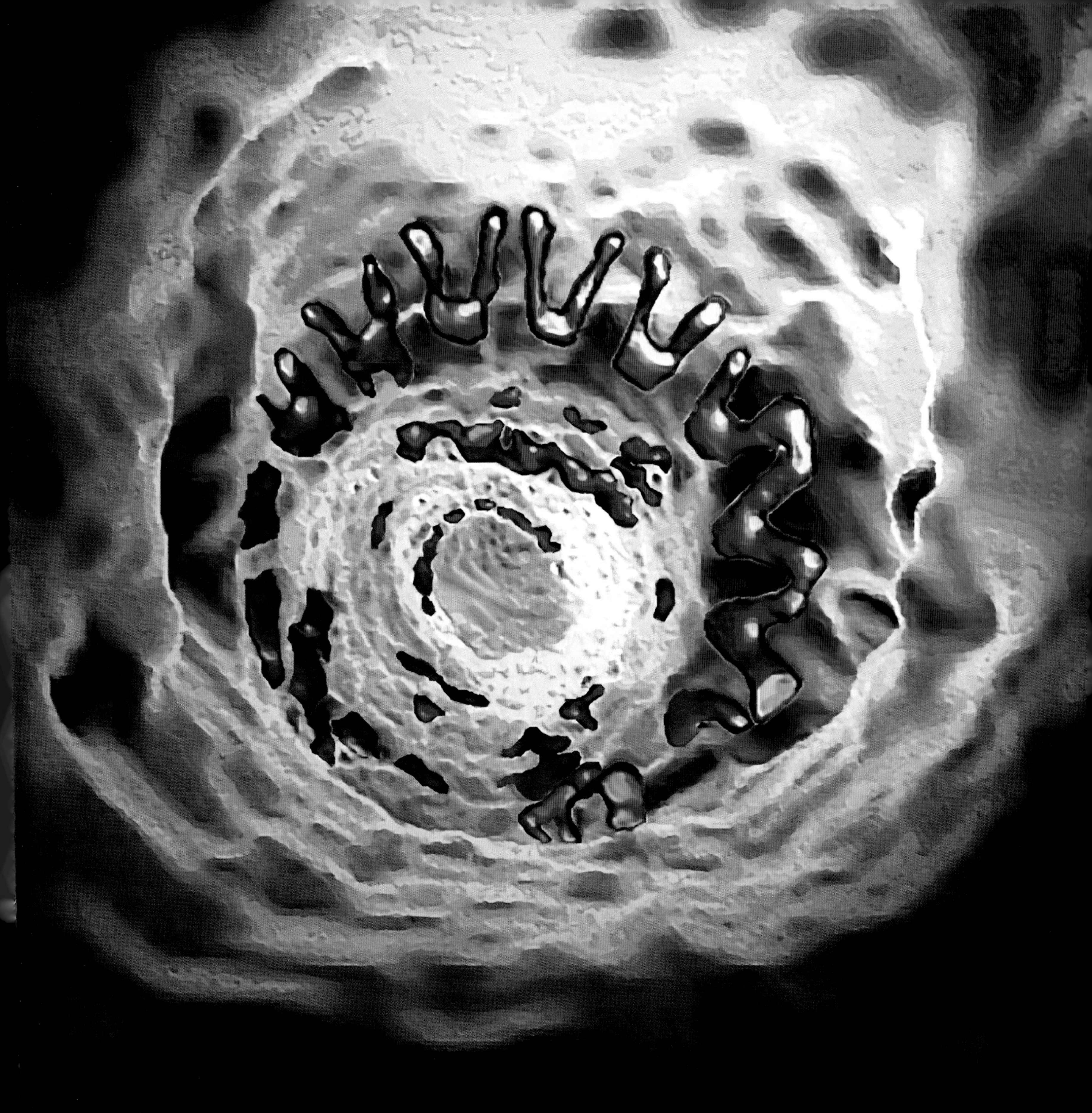

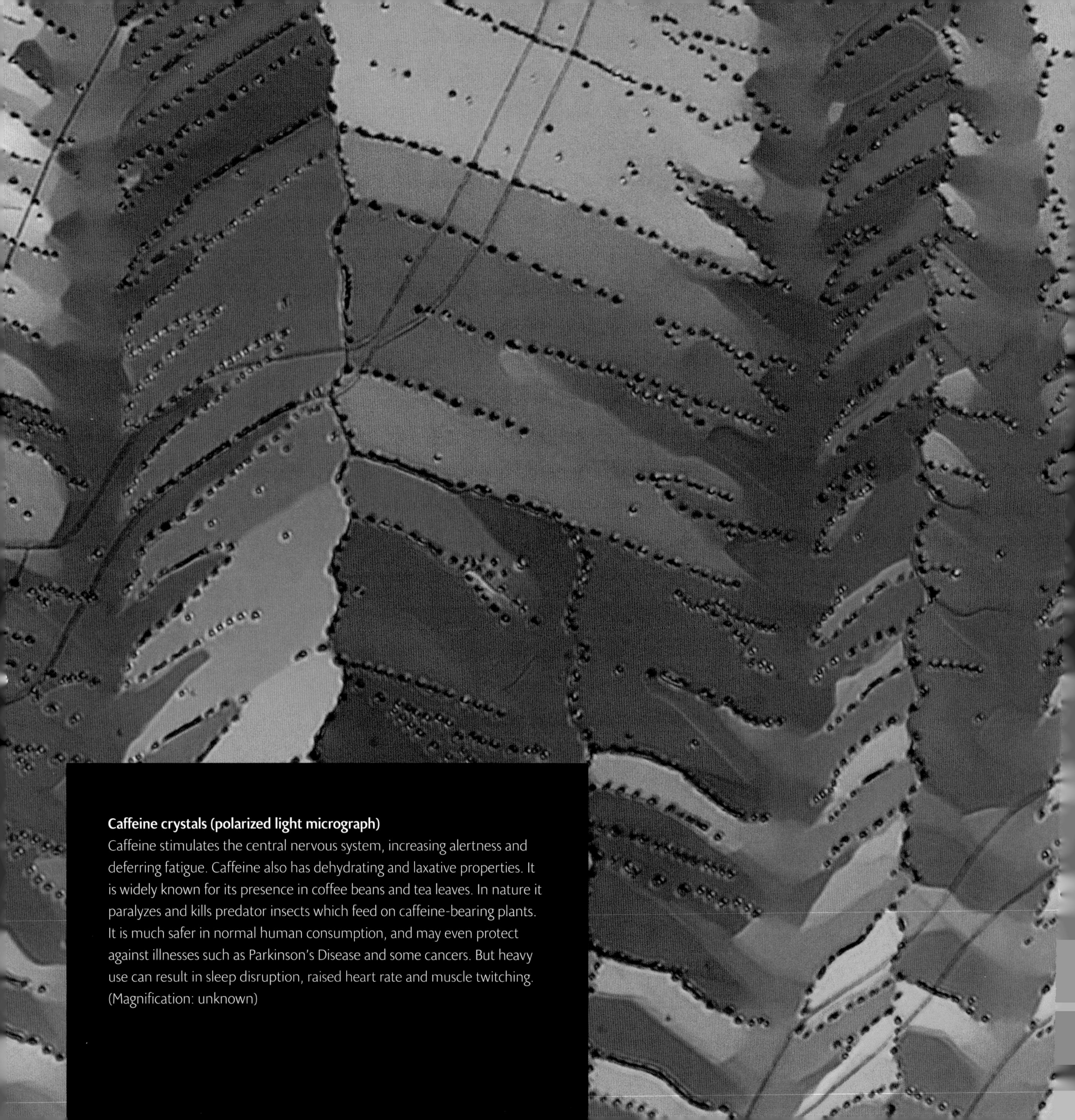

Caffeine crystals (polarized light micrograph)
Caffeine stimulates the central nervous system, increasing alertness and deferring fatigue. Caffeine also has dehydrating and laxative properties. It is widely known for its presence in coffee beans and tea leaves. In nature it paralyzes and kills predator insects which feed on caffeine-bearing plants. It is much safer in normal human consumption, and may even protect against illnesses such as Parkinson's Disease and some cancers. But heavy use can result in sleep disruption, raised heart rate and muscle twitching. (Magnification: unknown)

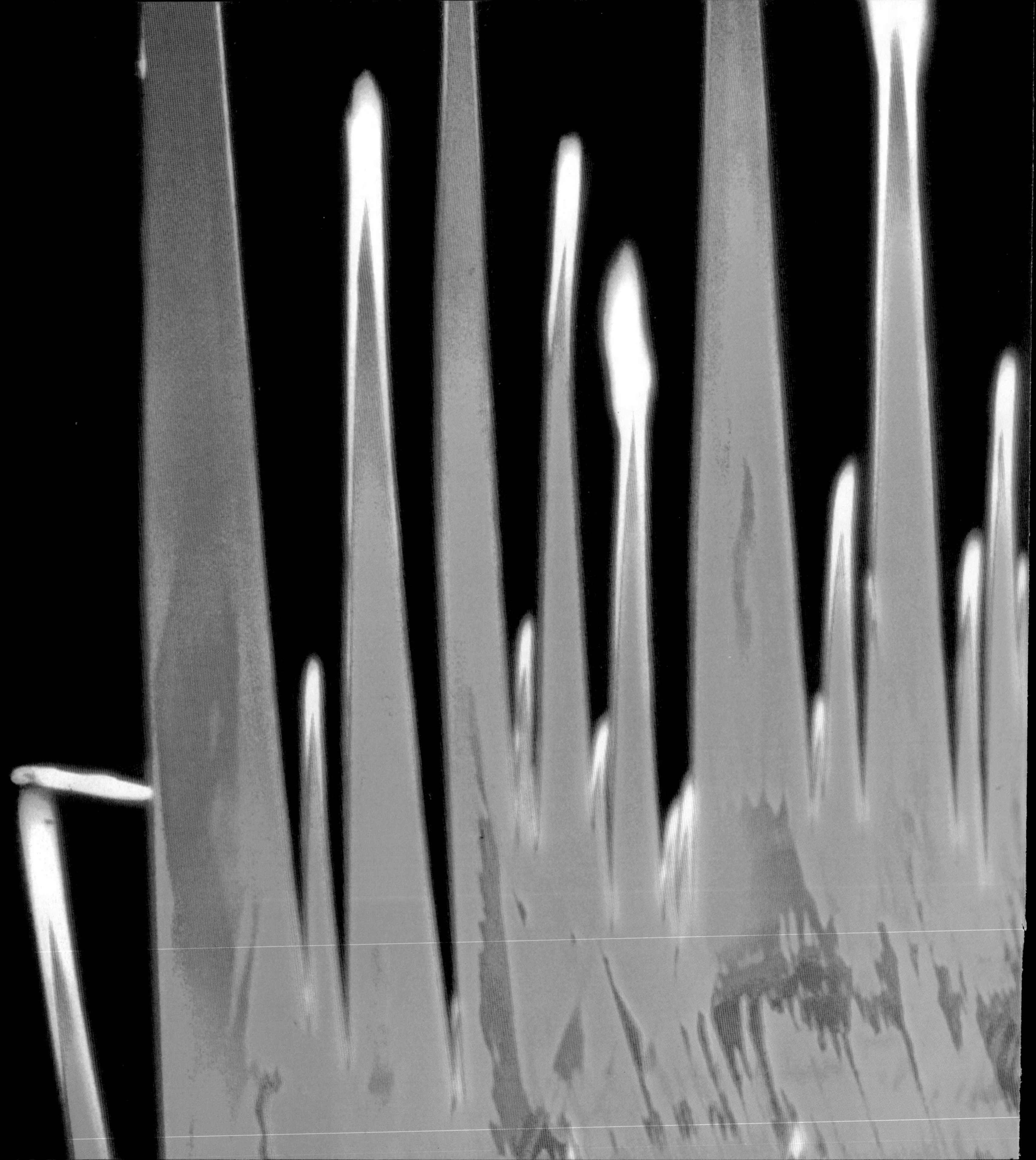

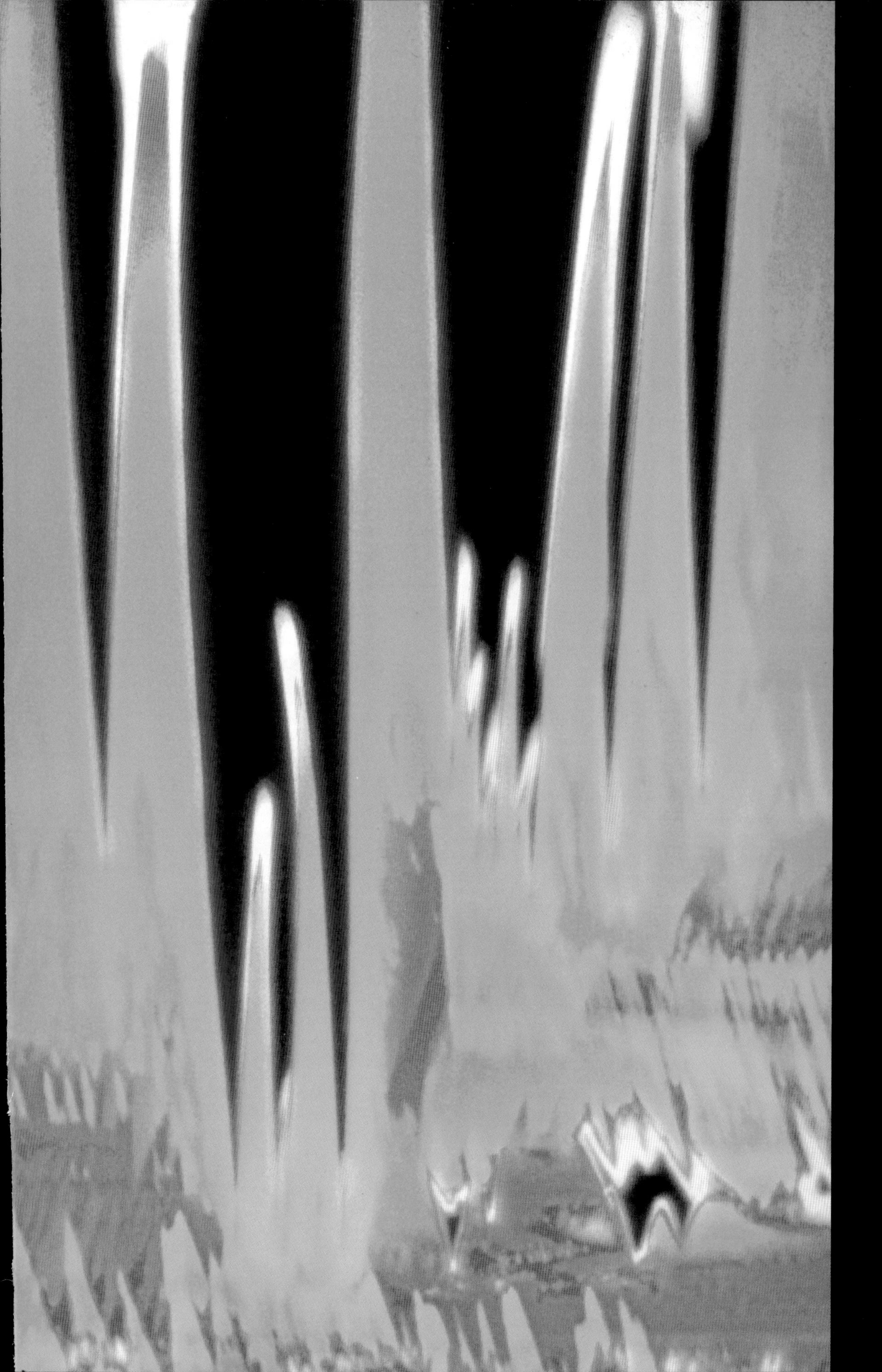

Folic acid crystals (polarized light micrograph)
Folate is a naturally occurring form of Vitamin B9. Its name is derived from the Latin word for 'leaf', and it is particularly abundant in dark green leafy vegetables. Folic acid is the synthetic form, found in supplements and fortified foods. Vitamin B9 is important for the production and maintenance of new cells. In the developing human foetus it prevents defects in the neural tube, which becomes the central nervous system. Spina bifida is the most common neural tube defect. (Magnification: x25 actual size)

Index

Picture Credits

Science Photo Library: 2 Antonio Romero, 12 Susumu Nishinaga, 13 Susumu Nishinaga, 14 Riccardo Cassiani-Ingoni, 15 Mike Agliolo, 16–17 Riccardo Cassiani-Ingoni, 18 Microscape, 19 Pasieka, 20 Dr. Torsten Wittmann, 21 Dr. Gopal Murti, 22–23 Thomas Deerinck, NCMIR, 24 Steve Gschmeissner, 25 Secchi-Lecaque/Roussel-Uclaf/CNRI, 26 Herve Conge, ISM, 27 Sriram Subramaniam/National Institutes of Health, 28 Carol N. Ibe and Eugene O. Major/National Institutes of Health, 29 C. J. Guerin, PhD, MRC Toxicology Unit, 30–31 Medimage, 32 Steve Gschmeissner, 33 Science Photo Library, 34 Science Photo Library, 35 L. Willatt, East Anglian Regional Genetics Service, 36–37 CNRI, 38–39 Dr. Stanley Flegler/Visuals Unlimited, 40 Steve Gschmeissner, 41 Steve Gschmeissner, 42 Steve Gschmeissner, 43 Steve Gschmeissner, 46 David Scharf, 47 Steve Gschmeissner, 48 NIBSC, 49 Science Photo Library, 50–51 Prof. P. Motta/Dept. of Anatomy/University 'La Sapienza', Rome, 52 Susumu Nishinaga, 53 Susumu Nishinaga, 54 Susumu Nishinaga, 55 Susumu Nishinaga, 56 Susumu Nishinaga, 57 Susumu Nishinaga, 58 Thomas Deerinck, NCMIR, 59 C. J. Guerin, PhD, MRC Toxicology Unit, 60 Biophoto Associates, 61 Prof. P. Motta/Dept. of Anatomy/University 'La Sapienza', Rome, 62 Science Photo Library, 63 Thomas Deerinck, NCMIR, 64–65 CNRI, 66 Riccardo Cassiani-Ingoni, 67 Medical Body Scans, 68–69 Pasieka, 70 Sherbrooke Connectivity Imaging Lab, 71 Pasieka, 72 Manfred Kage, 73 Steve Gschmeissner, 74–75 Steve Gschmeissner, 76 Prof. Stephen Waxman, Hank Morgan, 77 Manfred Kage, 78 Tom Barrick, Chris Clark, SGHMS, 79 Nancy Kedersha, 80–81 C. J. Guerin, PhD, MRC Toxicology Unit, 82 Steve Gschmeissner, 83 Michael Ross, 84–85 Michael W. Davidson, 86–87 Dr. Torsten Wittmann, 88 Prof. P. Motta/Dept. of Anatomy/University 'La Sapienza', Rome, 89 Prof. P. Motta/Dept. of Anatomy/University 'La Sapienza', Rome, 92 Pasieka, 93 Pasieka, 94–95 Michael W. Davidson, 96 Photo Indolite Realite & V. Gremet, 97 Susumu Nishinaga, 98 Steve Gschmeissner, 99 Louise Hughes, 100 Steve Gschmeissner, 101 Susumu Nishinaga, 102 R. Bick, B. Poindexter, UT Medical School, 103 Susumu Nishinaga, 104–105 Steve Gschmeissner, 106 Susumu Nishinaga, 107 Susumu Nishinaga, 108–109 Science Photo Library, 110 Pasieka, 111 Pasieka, 112 Science Photo Library, 113 Science Photo Library, 114 Steve Gschmeissner, 115 Dr. Keith Wheeler, 116 Prof. P. Motta/Dept. of Anatomy/University 'La Sapienza', Rome, 117 Susumu Nishinaga, 118–119 Dr. Keith Wheeler, 120–121 Pasieka, 122–123 Science Photo Library, 124 James Cavallini, 125 James Cavallini, 126 Science Photo Library, 127 Moredun Scientific Ltd., 128–129 James Cavallini, 130–131 Steve Gschmeissner, 132–133 Steve Gschmeissner, 134 Antonio Romero, 135 Steve Gschmeissner, 136 Steve Gschmeissner, 137 Steve Gschmeissner, 138 Steve Gschmeissner, 139 Steve Gschmeissner, 140 Clouds Hill Imaging Ltd., 141 Juergen Berger, 142 CDC, 143 Pasieka, 145 NIBSC, 146 Dr. Kari Lounatmaa, 147 Zephyr, 148–149 Centre Jean Perrin, 150 Dr. Dan Kalman, Katie Vicari, 151 NIBSC, 152 Pasieka, 153 Patrick Landmann, 154–155 Biozentrum, University of Basel, 156 Science Photo Library, 157 Moredun Animal Health Ltd., 158–159 Antonio Romero, 160–161 Michael W. Davidson, 162 National Institutes of Health, 163 National Institutes of Health, 164 Sidney Moulds, 165 Sidney Moulds, 166 Martin M. Rotker, 167 Eye of Science, 168 Steve Gschmeissner, 169 David McCarthy, 170 Thierry Berrod, Mona Lisa Production, 171 SCIMAT, 172–173 Steve Gschmeissner, 174 E. Gueho, 175 Photo Indolite Realite & V. Gremet, 176–177 Herve Conge, ISM, 178 Dirk Wiersma, 179 Leonard Lessin, 180–181 Michael W. Davidson, 182–183 Michael W. Davidson, 184 David McCarthy, 185 GJLP, 186–187 Dennis Kunkel Microscopy, Inc./Visuals Unlimited. **Corbis:** 5 Dr. Stanley Flegler/Visuals Unlimited, 10–11 SIU/Visuals Unlimited, 44–45 Dr. David Phillips/Visuals Unlimited, 144, 188–189 Dennis Kunkel Microscopy, Inc./Visuals Unlimited. **Getty Images:** 90–91 Garry DeLong.